Whiteness and Leisure

Leisure Studies in a Global Era

Titles include:

Karl Spracklen
WHITENESS AND LEISURE

Leisure Studies in a Global Era
Series Standing Order ISBN 978–1–137–31032–3 hardback
978–1–137–31033–0 paperback
(*outside North America only*)

You can receive future titles in this series as they are published by placing a standing order. Please contact your bookseller or, in case of difficulty, write to us at the address below with your name and address, the title of the series and the ISBN quoted above.

Customer Services Department, Macmillan Distribution Ltd, Houndmills, Basingstoke, Hampshire RG21 6XS, England

Also by Karl Spracklen

CONSTRUCTING LEISURE
HEAVY METAL FUNDAMETALISMS (*co-edited with Rosey Hill*)
SPORT AND CHALLENGES TO RACISM (*co-edited with Jonathan Long*)
THE MEANING AND PURPOSE OF LEISURE

Whiteness and Leisure

Karl Spracklen

Professor of Leisure Studies, Leeds Metropolitan University, UK

palgrave
macmillan

First published 2013 by
PALGRAVE MACMILLAN

Palgrave Macmillan in the UK is an imprint of Macmillan Publishers Limited, registered in England, company number 785998, of Houndmills, Basingstoke, Hampshire RG21 6XS.

Palgrave Macmillan in the US is a division of St Martin's Press LLC, 175 Fifth Avenue, New York, NY 10010.

Palgrave Macmillan is the global academic imprint of the above companies and has companies and representatives throughout the world.

Palgrave® and Macmillan® are registered trademarks in the United States, the United Kingdom, Europe and other countries.

ISBN 978–1–137–02669–9

This book is printed on paper suitable for recycling and made from fully managed and sustained forest sources. Logging, pulping and manufacturing processes are expected to conform to the environmental regulations of the country of origin.

A catalogue record for this book is available from the British Library.

A catalog record for this book is available from the Library of Congress.

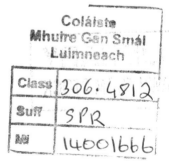

For Beverley

Contents

Series Preface

In this book series, we defend leisure as a meaningful, theoretical, framing concept; and critical studies of leisure as a worthwhile intellectual and pedagogical activity. This is what makes this book series distinctive: we want to enhance the discipline of leisure studies and open it up to a richer range of ideas; and, conversely, we want sociology, cultural geographies and other social sciences and humanities to open up to engaging with critical and rigorous arguments from leisure studies. Getting beyond concerns about the grand project of leisure, we will use the series to demonstrate that leisure theory is central to understanding wider debates about identity, postmodernity and globalization in contemporary societies across the world. The series combines the search for local, qualitatively rich accounts of everyday leisure with the international reach of debates in politics, leisure and social and cultural theory. In doing this, we will show that critical studies of leisure can and should continue to play a central role in understanding society. The scope will be global, striving to be truly international and truly diverse in the range of authors and topics.

Professor Karl Spracklen and Professor Karen Fox

Acknowledgements

Thanks as always to everybody at Palgrave Macmillan.

1
Introduction: Thinking about the Problem

It could be argued that a monograph on whiteness and leisure is irrelevant or dangerous: irrelevant because the world has changed, all our identities are liquid and all social structures have melted away; dangerous because such a monograph has the potential to essentialize racial identities and – worse – recreate hierarchies of belonging based on fixed ontological categories of 'race' or ethnicity. This monograph does not essentialize whiteness, nor does it simply reproduce fixed notions of identity. Whiteness is always being constructed, challenged and re-defined. This book shows how whiteness and contestations of whiteness and Otherness are (re)produced in and through leisure: how 'race' is a problematic ontological category. However, that is not to say such categories are irrelevant. This book is timely because leisure is a form and space where inequalities of power are refracted through social structures and material and cultural power is at work making constructions of whiteness unproblematic. This book's aim is to shine a light on this activity.

Consider two news stories that circulated in the United Kingdom at the end of 2011, when I first drafted this introductory chapter. The first was a headline story in the sports sections of newspapers about racial abuse allegedly made by the England football (soccer) captain John Terry towards an opposing (black) player in a Premier League match ('Ferdinand to be questioned by FA over Terry affair today', Stuart James, *The Guardian*, sports section, p. 2, 28 October 2011). The big question was whether Terry had actually used racist phrases, and whether this allegation would deprive him of his England captaincy. Terry claimed he was actually telling the black player he had not said anything abusive – and the television cameras caught him at the moment he said 'black ****' but missed the words 'I didn't call you a . . .', which supposedly prefaced the racial abuse. The truth of the matter came out over the next 12 months, and although Terry was found guilty by the Football Association, he received the most lenient ban possible ('John Terry verdict: if the FA does not think this is racism, what is?', David Conn, *The Guardian*, online at http://www.guardian.co.uk/football/2012/oct/05/john-terry-fa-commission, 5 October 2012, accessed 3 December 2012). At the point in time I first wrote this paragraph, though, Terry's whiteness had not been the subject of any debate in the newspapers: he was just Terry, the England captain, the working-class boy 'done good', the hero on the poster of a million bedroom walls. Terry was white, and represented a form of working-class whiteness, but that whiteness was never a problem; he was never the victim of racial abuse about his whiteness.

The second news story was a satirical report about the celebrity 'medium' Derek Acorah ('Stop me if you've heard Derek's predictions before', Marina Hyde, Lost in Showbiz, *The Guardian G2*, pp. 2–3, 28 October 2011). Acorah, like all psychics, is very probably a fake who makes things up or finds things out so that he can impress his paying audiences, who then pay him again the next time he comes round (it is quite simple to spot the tricks of the professional fraud and quite easy to develop those tricks oneself without any need for supernatural intervention – see Rowland, 2002). What is of interest to the theme of this monograph is the reporting of Acorah's show in the northern English town where he was allegedly contacted by the dead spirit of someone called George. It is a safe bet that somebody in the audience had a dead relative or friend called George, and somebody claimed George as their sorely departed. Acorah then claimed that George 'still doesn't like people from other countries, especially the darker people' (Hyde, 'Stop me . . .', p. 3). He, however, did finish by saying that George's comment is not very nice. Now it could be that Acorah was psychic and George was just a racist ghost. But it is almost certain that Acorah was not psychic and George was made up by Acorah: to pander perhaps to the old white English people in his audience; to make George realistic by having him express views old white English people (or white fake psychics) hold; or more likely both of those things. The leisure activity of cabaret theatre then – or its psychic medium subculture – was perceived as a white space, for the use and entertainment of white people. Acorah can distance himself from his ghost's views, but Acorah is part of the construction of the whiteness of his audience and – like a good fake – he is also good at pointing us away from this construction.

His further claim that he has an African spirit guide again exoticizes blackness and makes it an acceptably foreign thing completely alien to his unmentioned and invisible whiteness.

Both these stories demonstrate the work of 'race' in contemporary British culture, leisure and sport: making blackness some Other outside of mainstream, white Britishness and hiding the problematic social construction of whiteness. In this monograph, I will show how this work is done across leisure, and across the modern West, from the United States to Australia. It is necessary at this point to stress the unreality of 'race'.

* * *

In 2008 I published a research note about beliefs in the biological nature of 'race' among sport scientists. In the discussion, I made it clear why such beliefs were scientifically and philosophically wrong (Spracklen, 2008: p. 225):

> The problem with 'race' as a category is the movement of people in the last 400 years (through colonisation, commerce, slavery), and especially in the last 100 years (through globalisation, industrialisation and migration), has made racial categories impossible to sustain in any useful or meaningful sense (Banton, 1998). There are no discernible genetic differences between 'black' people and 'white' people (Franks, 2007). Phenomes (e.g. the ability to be an elite sprinter) cannot be mapped on a one-on-one basis onto genomes (genetics), so there cannot be a causal link associated with heredity (Hoberman, 2004). That is not to deny that there are clusters of populations that are more or less likely to be carriers of particular genetic information, but the existence of such clusters is not the basis for an ontology of racial difference (Skinner, 2006). Indeed, the caution with which claims about particular populations are made suggests that such clusters are dynamic, partial and rare. The burden of proof has to be on those who make claims of racial difference (Shim, 2005). What is happening is a category error: scientists assume races exist because the myth of the Holy Blood makes 'race' normal and unproblematic, and experiments are designed on that basis. Hence the gobbledegook of claiming, as in the fast-twitch muscle fibre experiments cited by Entine (2000), that Afro-Americans (a diverse group) are defined as West Africans. What Entine is actually showing is that most successful sprinters are American, and the best sprinters have more fast-twitch fibre.

What these sport scientists were doing was essentializing the notion of blackness, turning a social phenomenon into a biological category. In doing this, they ignored the social construction of blackness, the particular histories of black people in the United States and the rest of the West, and the ways in which certain sports became spaces for the construction of blackness and black masculinities (Carrington, 2010a). But the sport scientists were also missing out the social construction of whiteness in their myth-making about black physicality. First of all, there were no claims about the biological nature of white people and how such natures might explain in a *post hoc* way the dominance of white people in certain elite sports events. Secondly, and more important, there was no attempt to account for the sports where opportunities for black people's participation was limited or denied altogether through the unwritten rules of belonging and exclusion (Bourdieu, 1986; Long and Spracklen, 2010).

Put another way, the way in which sport is used to construct whiteness, and the way in which whiteness shapes sports, is an important unanswered theme in the sociological critiques of sport, 'race' and racism (Hylton, 2009; Long and Spracklen, 2010). This gap in the theory and the empirical research is also seen when the focus of the myth-making – sport – is expanded to include the entire range of people's leisure lives, leisure activities and leisure spaces. This monograph is an attempt to address this gap by developing a new theory of whiteness and leisure, which draws in part on existing leisure theories and in part on the critical theorizing around 'race' and whiteness associated with Critical Race Theory (CRT) and other radical social theories. In developing a new theory of whiteness and leisure, new primary and existing secondary empirical research will be drawn upon to highlight whiteness across a comprehensive and internationally grounded range of leisure practices. This monograph will analyse sports participation, sports media and sports fandom, but it will also analyse informal leisure, outdoor leisure, music, popular culture and tourism. This will make this monograph unique: an essential introduction to whiteness and leisure; an important development of leisure theory; a critical analysis of leisure practices, which will include new primary research on the construction of whiteness in a number of sport, leisure, tourism and popular culture activities; and a contribution to the critical theory literature on whiteness and 'race'.

This monograph book is grounded in leisure theory, and in particular my own development of leisure theory (Spracklen, 2009, 2011) in applying a Habermasian framework of communicative and instrumental rationalities and actions (see Habermas, 1984, 1987) to understanding the tensions between utopian theories of individualized, postmodern leisure (Blackshaw, 2010; Rojek, 2010) and dystopian theories of increasing constraint and control (Bramham, 2006). This third part of a trilogy I started with *The Meaning and Purpose of Leisure* and continued with *Constructing Leisure* will again look to Habermas' insights into the interaction between communicative reason and instrumentality to situate different whitenesses in broader political structures. What I am interested in here is the way in which leisure choices are used to construct exclusive, white identities – whiteness associated with individualism and elitism but also subordinate whitenesses that do the political work of the elites while being hegemonically constrained. I am interested in the 'beating of the boundaries' (Appelrouth, 2011; Cohen, 1985) – who is allowed to define belonging in various leisure activities and leisure spaces, from the subcultural scenes of pop music through everyday leisure lives and tourism to sports.

Whiteness throughout this book is used to represent a particular, hegemonic but invisible power relation that privileges (and normalizes) the culture and position of white people (Daynes and Lee, 2008; Dyer, 1997; Garner, 2006; Gilroy, 2000; Long and Hylton, 2002). The whiteness of white people can never be essentialized – there is no such thing as a white race and there is no such thing as a black race (Daynes and Lee, 2008). However, blackness and whiteness, the agency of choosing to identify with one or the other and the instrumentality of defining those who do not belong to one or the other (the Other, as it were) are part of what Daynes and Lee (2008) call the 'racial ensemble', tools used in boundary work, the formation of cultural capital (Bourdieu, 1986) through communicative agency and instrumentalized consumption. Where whiteness differs from blackness is in its link to the dominant side in historical inequalities of power and the useful instrumentality of universalizing white cultural norms as universal norms. In leisure, blackness is inevitably Othered as exotic, and the whiteness of everyday leisure forms is made invisible (Hylton, 2009; Long and Hylton, 2002; Long and Spracklen, 2010). Although the focus of the book is whiteness, it will be impossible to discuss 'race' without discussing the intersectionality of 'race', class, gender and sexuality. Throughout the book, whiteness will be examined through this intersectional lens – and intersectionality will be returned to in more detail in the Conclusions (Chapter 12).

The rest of the book

Chapter 2 will provide a clear and coherent review of literature on critical theory of 'race' and whiteness. This chapter will begin by examining different theoretical frameworks about 'race' and identity from sociology and cultural studies, focussing in particular on the work of Stuart Hall, Paul Gilroy and the main advocates of CRT. It will provide an overview of the key concepts, relationships and tensions. The section titled 'Theories of whiteness' will focus on theories of whiteness, white privilege and white power, drawing on the metaphor of the white mask used by Franz Fanon to begin to understand the ways in which whiteness is normalized in modern, Western society. In both sections, I will criticize scholars for obscuring their analyses with over-theorized language, and I will identify and stress a new approach to understanding 'race' that engages readers with clarity: this commitment to a clear style will be taken on through the remainder of the book. The two sections of the chapter will be brought together in a critique of contemporary culture and politics, and the importance of leisure as a site of hegemonic, white control and counter-hegemonic, racialized resistance will be shown to be linked to leisure's importance in modern life.

Chapter 3 will provide a review of research in sport, leisure, tourism and popular culture that discusses whiteness. The chapter is arranged into four sections. The Section 'Whiteness in leisure studies' will be a critical discussion of research across the broad remit of leisure studies that explicitly engages with whiteness. Specific research examples from sport, leisure, tourism and popular culture will be considered in this section in more detail. The following section 'Blackness in leisure studies' will critically analyse research from leisure studies that has a focus on blackness, but where whiteness is implied or invoked without any consistent attempt to problematize the concept. I will argue in this chapter that leisure studies, while providing a strong set of research examples that demonstrate the whiteness of leisure, are yet to provide a coherent account of whiteness and leisure.

Chapter 4 will provide a new theory of whiteness based on Habermas' insights into communicative reason and instrumentality. I will first discuss Habermas' contribution to our understanding of social identity and power, positioning his work in wider critical theory. I will then discuss criticisms of Habermas from the post-structuralist school, which position Habermas' defence of the Enlightenment project as a retreat into Eurocentric discourses of white, male power. I will show that such criticisms are philosophically naïve and suffer from a self-contradiction in their argument. Rather than reifying white privilege, Habermas' defence of communicative reason provides a space in which such privilege can be challenged – when such privilege becomes identified with a form of instrumental rationality about 'race'. Whiteness becomes an all-pervasive instrumentality, which, like capitalism, threatens to consume the entire world. The existence and survival in the Academy of counter-narratives of 'race', predicated on communicative rationality, shows that the Enlightenment – while flawed in history – remains a durable ideal of free inquiry.

Chapter 5 will focus on whiteness and popular culture. The chapter will begin with an introductory section, which provides a secondary analysis of the existing research literature on whiteness and popular culture. The rest of the chapter is divided into four sections, which will explore whiteness in particular contexts of popular culture and its relation to leisure. The second section of the chapter will focus on whiteness in television, drawing on and using examples from popular American TV shows syndicated worldwide, such as *Friends* and *Star Trek*. It will examine how leisure is used to construct whiteness within the shows and how watching the shows teaches consumers to be white. The third section of the chapter will look at whiteness and fantasy films and online gaming platforms such as World of Warcraft. It will show how these films, gaming platforms and video gaming can be used to counter instrumental whiteness by allowing some communicative space to subvert and resist dominant discourses – but such communicative spaces are rarely used by fans and gamers. The fourth section of the chapter, and the third example of whiteness and popular culture, will be an exploration of popular literature: glossy magazines aimed at mass markets (male and female) and best-selling books (genre novels and non-fiction lists). It will be shown that instrumental whiteness, like hegemonic masculinity, is fully dominant in this part of popular culture, and leisure choices inevitably reproduce such whiteness at the expense of other identities.

Chapter 6 will focus on whiteness and music. The chapter begins with an introductory section followed by a second section that will explore the distinctions between classical and pop, and rock and rap/R&B, to identify the racialized discourses present in music. The next section of the chapter will then proceed to explore world music and roots music, and different notions of whiteness and Otherness present in discussions of authenticity in these genres. The fourth section of the chapter will introduce new primary research by the author on whiteness and nationalism in English, European and American folk music and European black metal, two forms of music unrelated by sound but with a shared susceptibility to infiltration by far-right nationalists. The concluding section will argue that those who listen to or play music in their leisure lives can make communicative choices that resist instrumental whiteness, but the commodification of most of music industry makes this incredibly difficult and leads to the real danger of unwittingly supporting racialized discourses.

Chapter 7 will focus on whiteness and sport and will draw on secondary analysis of sociology of sport research, along with some primary research on the history of modern sports and contemporary sports. The chapter will draw on examples from the United States, Australia, South Africa and Europe. The section 'Sports participation' will examine sports participation and the construction of cultural capital through involvement in sport. Whiteness will be identified as an invisible, taken-for-granted signifier in modern sports, with participation among white people in certain sports such as basketball and athletics being dependent on sports that provide safe 'white spaces'. I will demonstrate that when significant numbers of black people start to take an interest in these sports, white people start to choose other sports to play. The following section titled 'Sports fandom' will focus on sports fandom and the modern, professional sports industry. I will show that white people find an imagined community of 'pure' whiteness in supporting a particular club or national sports team (for example, the South African Springboks) and this whiteness is only partially challenged by the introduction of black athletes into those teams. White sports fans, I will argue, are (generally, mostly) comfortable with black athletes representing their club or their sport or their country because the globalization and professionalization of sport has turned young black athletes into caricatures of physicality and modern-day slaves. White sports administrators are still in charge of the sports, still defending the white history of these sports even as some black athletes are embraced as exotic 'ringers'. As such, the power dynamics ensure whiteness remains hegemonic.

Chapter 8 will focus on whiteness and sports media. This is a companion chapter to Chapter 7. It could be argued that this chapter should be subsumed into the previous one, but 'sports media' is a completely different leisure category to sport: most people who watch sports on television or read about them on the Internet do not watch sports events live, and do not participate. This chapter discusses the way in which sports and entertainment intersect and construct whiteness. I will begin this chapter by demonstrating that sports media are one part of a wider entertainment industry, globalized, commodified and controlled by a small number of trans-national corporations, which in turn have a close relationship with the anti-regulation, pro-liberalization trends of modern political parties. Sports play a key role in increasing the profits of such corporations, so sports media have always been exploited by such corporations since the advent of radio in the first half of the twentieth century. This chapter will proceed to discuss examples from Australasia, North America and Europe of recent and current sports media and the way in which such media over-signifies blackness and makes whiteness ordinary and invisible. I will show that sports media – whether traditional television programming of live sports events, or modern media outlets such as online blogs and discussion forums – are places where instrumental, hegemonic whiteness is constructed, with little room for any communicative resistance from subaltern groups.

Chapter 9 will focus on whiteness and everyday leisure. This chapter will involve both secondary analysis of existing research and some primary research around everyday leisure activities such as shopping, eating out, drinking, dancing, reading and interacting on social networks. The chapter will span the Western world, with examples drawn from everyday leisure in North America, South Africa, Australia and Europe. In the section 'Everyday leisure in the west' I suggest that everyday leisure activities are particularly problematic precisely because they are viewed as everyday mundane, normal, routine leisure choices. None of these everyday leisure activities are necessarily white: however, the whiteness of these activities is (re)produced in the cultural capital these activities construe. In the section titled 'Ethnic food' I focus on the phenomenon of 'ethnic food', drawing on existing research and new data collection to examine the ways in which eating such food constructs both a foreign Other and a normalized, white Us.

Chapter 10 will focus on whiteness and tourism. This chapter will focus on three areas of tourism where there is existing research to enable a detailed secondary analysis: package holidays in Europe and the Far East; heritage tourism in the United States and the United Kingdom; and independent travelling along well-used routes from the West (and Australia) to the East and South America. The chapter will begin with a short overview of the history of the modern tourist industry and the assumptions about whiteness and (post)imperialism implicit in the concept of travelling to another place 'on holiday'. The section on package holidays will discuss subordinate whitenesses based on working-class Western identities and identify a residual cultural resistance to the norms of hegemonic, white, middle-class Westernness. The section on heritage tourism will discuss the ways in which dominant whitenesses are imagined, and the instrumental nature of that (re)imagining, which privileges elitist narratives of place and nation. The final section on the subcultural identity of the traveller will explore the tensions between traveller beliefs about cultural diversity and the white neo-imperialism and colonialism such travellers bring with them to the countries they visit.

Chapter 11 will focus on whiteness and outdoor leisure: leisure activities such as walking, climbing, cycling, canoeing, sailing, skiing, cross-country running and involvement in conservation volunteering. The second section of the chapter will survey the research literature on outdoor leisure to identify the presence or absence of whiteness in the theoretical frameworks used to understand outdoor leisure. The rest of the chapter will focus on long-distance walking in North America and the United Kingdom, and will use new primary research to show how such outdoor leisure practices are connected to the imagined, nationalist nature of instrumental whiteness.

Finally, Chapter 12, the concluding chapter, will provide a synthesis of the key findings of the rest of the book and return to the problems raised in the first three chapters to show that the seemingly inevitable relationship between modern leisure and whiteness (and whiteness and intersections of class and gender) is not a necessary feature of modern life; it is only a contingent one.

Throughout the book I will draw on my own experiences, reflections and recollections of my own leisure – as a white, Western man. I will draw on informal discussions that I have recorded or recalled and things that I have come across on the Internet and in the press. Where I am reflecting I will use text written in italics to mark the change in tone.

2
Theories of 'Race' and Whiteness

Introduction

The history of 'race' is deep-rooted. All communities and cultures identify insiders, members of the group and outsiders, those who are categorized as the Other: the unknown, the alien, the strange and the exotic. It is quite possible to imagine a world where history ran differently, and people who defined themselves as greens enslaved purple people and considered this was because something about their green identity made them innately superior to purples. The greens might see this superiority as a fact of their breeding, or their culture, or their divinely written destiny. They might claim that purple people are weak, base animals, not truly human. This is just a silly fantasy story, a thought experiment that tries to approach 'race' and racism from a place abstracted from real life. But already what I have written resonates with the actual history of our world. How we think about 'race', racism and power today is predicated on an understanding of the emergence of Western power in the Early Modern period, the shaping of European empires, the growth of slavery in the Atlantic, the spread of Enlightenment thinking, the Industrial Revolution, the passage of individuals around the world, the rise to dominance of the West and the United States, the abolishment of slavery, the globalization of culture and modernity, post-colonial shifts of power and the spread of global capitalism.

The origins of modern racialized categories are found in two of the most important cultures that shaped the thinking of the Enlightenment philosophers who in turn shaped Western modernity: the classical world of the Romans and Greeks, and the Christendom of medieval Europe that emerged in the West following the slow death of the Western

Roman Empire. For the Greeks and Romans, their peoples were defined not only by blood and belonging (citizenship being something inherited through family ties, being marked out by geographical spaces and physical communities sharing supposedly common origins) but also by civilization (being literate, living in cities), against the barbarians beyond the limits of civilization. The Greeks and Romans considered their civilization to be superior to the wild, savage natures of barbarian life – and this sense of civilization being superior to the barbarian Other is key to understanding the social construction of 'race' and racism in modernity. Westerners took it as read that Western civilization was good because it was their civilization, inherited from the Romans and Greeks who dominated their ruling classes' education, and the political and economic successes of the eighteenth and nineteenth centuries seemed to demonstrate the superiority against the savage 'barbarians' the Westerners encountered. Medieval Christianity did bring a conception of a Biblically inspired common humanity, but this version of Christianity also brought with it the idea of a 'Chosen People' grafted to pre-Christian ideas of blood ties, belonging and native soil. This concept shaped the worldview of the Protestants who shaped the emergence of modern science, modern capitalism, the British Empire and the United States. It was also prevalent in Catholic countries grappling with the growth of the public sphere, the shift from feudalism to modernity, and the rise of the nation-state. If a given country was chosen by God, who belonged to the Chosen People? What did it mean to be English, or British, or French, or American? By the nineteenth century, such questions were increasingly answered by reference to the existence of a particular 'race' or races (Aryan, Anglo-Saxon, Nordic, European, Northern European or white) that was in nature and in culture superior to its rivals. Those rivals were in turn investigated and categorized using the scientific methods and assumptions of the day: just as scientists were able to define and delineate the nature of elements of the Periodic Table, so they argued, they could categorize the biological differences that marked out one 'race' from another. Racial science was riven by disputes over how many races existed, and where one drew the line between different races (a dispute, of course, that was caused by the fact that the races did not exist as biologically discrete entities, fixed by nature). At the same time, anthropologists started to describe and discuss the cultural differences between the supposed races, and although some anthropologists stressed the universal nature of human cultural practices, many took the language of 'race' and applied it to their own research to try to demonstrate that the fixed nature of cultural difference was just like the Romans and Greeks had described it: that is, there were clever, civilized, urbanized people, versus barbarians fixed by their cultural habits.

To understand the social construction of whiteness in leisure and modern, Western society, it is necessary to consider the wider debates about the meaning and purpose of 'race' and the history of critical thinking about 'race' and racism. The purpose of this chapter is to provide a clear and coherent review of literature on critical theory of 'race' and whiteness. Obviously, in doing this I can only focus on key issues, theories and theorists in this vast literature. This chapter will begin by examining different theoretical frameworks about 'race' and identity from sociology and cultural studies, focussing in particular on the work of Stuart Hall, Paul Gilroy and the main advocates of Critical Race Theory (CRT). It will provide an overview of the key concepts, relationships and tensions. The second section of the chapter will focus on theories of whiteness, white privilege and white power, drawing on the metaphor of the white mask used by Franz Fanon to begin to understand the ways in which whiteness is normalized in modern, Western society. In both sections, I will criticize scholars for obscuring their analyses with over-theorized language, and I will identify and stress a new approach to understanding 'race' that engages readers with clarity: this commitment to a clear style will be taken on through the remainder of the book. The two sections of the chapter will be brought together in a critique of contemporary culture and politics, and the importance of leisure as a site of hegemonic, white control and counter-hegemonic, racialized resistance will be shown to be linked to leisure's importance in modern life.

The sociology of 'race' and racism

American scholars started to theorize 'race' and racism in the first half of the twentieth century. These scholars were interested in the 'problem' of black poverty and the prevalence of racial prejudice. The particular history of the United States, in particular the long history of slavery and genocide of Native Americans, coloured sociological accounts of the inequalities facing black Americans. For some white American scholars, the inequalities facing black Americans were a consequence of their race: following Social Darwinist theories, they argued that black Americans were too lazy or too stupid to free themselves from poverty and become 'normal' citizens (see discussion in Kohn, 1995). Such racist ideology was also applied to recent migrants into the country, and also to the Native Americans who still survived. For the Social Darwinists, the failure of non-white people to have successful lives in modern capitalism was a fault not only of their blood and their biological essence but also of their dangerously immoral cultural habits. Such racist ideology had a long history in the American academy, and some of that ideology survives in popular scientific notions about 'race' and in disciplines such as physical anthropology and sports science (Spracklen, 2008). Such ideology, however, has been challenged by a series of critical theorists, from Du Bois (1986) all the way through to Omi and Winant (1986), who have all argued that black Americans and other ethnic groups within the country have been systematically marginalized and excluded from having power and access to the mainstream of modern society. Critical theorists of racism have suggested that racial hierarchies are imposed on black people and other individuals, who are then identified as outsiders and problem citizens (Daynes and Lee, 2008). For some American theorists of 'race', the marginalization of blackness can only be defeated by challenging such processes through a re-evaluation of the positive nature of Afro-American history and culture (Asante, 2001). This has led full circle to a politics of blackness that subscribes to Afrocentric ideology and its own essentializing racial myths about the purity of blackness (Daynes and Lee, 2008).

In Europe, the sociology of 'race' and racism is mostly influenced by the history of European empires and the post-colonial settlements, in which empires collapsed and millions of people moved out of and into European nation-states as economic and political migrants. Marxism and post-Marxism shape the thinking of scholars of 'race' such as Richard Miles, Stuart Hall and Paul Gilroy. For Miles (1989), racism is a tool of modern capitalism, used to split the working classes and turn them against each other. Racial exclusion is for him another structural exclusion, like gender, that operates to perpetuate the power of capitalism and the modern nation-state against those whose labour is exploited. In the contemporary history of Europe, one can see the ways in which economic migrants are exploited by capitalism while they are also stereotyped as scapegoats by the right-wing, capitalist press. Racial stereotyping and racism for Miles will disappear if the current unequal power relationships of capitalism are replaced with more equal forms of distribution. While this Marxist framework is useful in identifying the fact that 'race' is in many ways reducible to class, there is an absence of identity politics and culture politics in its analysis. Stuart Hall is the European scholar who is best known for attempting to analyse 'race' and racism through identity and culture. Hall (1993, 1995, 1996, 1997, 2002) argues that every individual in late and post-modernity tries to make sense of the world around them – and their place in that world. We create our identity and label others through recourse to the cultural, social and political tools we have to hand. In modern societies, says Hall, racialized identities are markers not only of subjugation and control but also of resistance: blackness can become a way of fighting against the inequities of modern capitalism (Hall, 1995). In postmodernity such cultural identity-working becomes more fluid and hybrid, but because of the reduced structures that bind such societies, racialized identities become a place of playful creation and potential political action (Hall, 2002). In *The Black Atlantic*, Paul Gilroy (1992) identifies the symbolic community of the black diaspora, the 'Black Atlantic' of the book's title, which through the circumstances of history binds together black communities in America, the Caribbean and Europe. This diasporic community shares cultural practices such as music and food, but it is also hybrid, belonging in part to the countries in which particular black people live. For Gilroy this is a key strength of blackness, but there is a danger for him in promoting culture too much at the expense of challenging essentialism (see also Gilroy, 1987). In his later work, Gilroy (2000) has critiqued identity politics for allowing and legitimizing essentialist discourses back into the academic and political arenas: if blackness is so different, then it allows white people off the hook for the inequality and exclusion black people have suffered. Caution should be exercised in the celebration of racialized identities so that it does not become a way of maintaining the unequal power relationships of modern capitalism: it is right to celebrate blackness through the funding of particular events such as Black History Month, for example, because such history has been marginalized ('white-washed') until very recently as a result of the racism or unconscious prejudice of earlier Western historians; but it is not right to support white supremacists in creating a White History Month, which would only retell the history that has been told before, the story of the people in power, twisted into an even more egregiously racist ideology. This cautionary tone is picked up in the work of theorists who highlight the intersectional nature of our identities: all identities are contingent, all are in a process of becoming, but some are marginalized through the circumstances of political history. So gender, class, 'race' and ethnicity and sexuality intersect with one another, creating further exclusion and allowing those with power to create stereotypes about outsiders (Nayak, 2003, 2006).

Such stereotypes are at the centre of 'race' and racism, especially the ways in which we think about 'race' and leisure. Gilman suggests (1985) that stereotypes are a necessary, and in a way inevitable, part of human social interaction, as they help us to make sense of the world from the earliest moments of childhood. However, it should be remembered that stereotypes are not neutral vehicles by which we make sense of the 'real world' around us. We attach inferior and superior values to them and they are historical formations that have an inherent power dimension. In this sense it can be argued that stereotypes are, by their very nature, linked to questions of control and have particular consequences because of asymmetrical power relations between groups. As Fiske (1993, p. 623) argues:

> Stereotypes control people, which is one reason why they are so aversive. No one wants to be stereotyped. Stereotypes reinforce one group's or individual's power over another by limiting the options of the stereotyped group, so in this way stereotypes maintain power. People with power do not have to put up with them, but people with-out power are victims. Power is control, and stereotypes are one way to exert control, both social and personal.

The 'social effects' of such practices can be very damaging. The power of the stereotype extends to the group that is being stereotyped; because of the asymmetrical power relations, their ability to define themselves is constrained by the dominant discourse (see, for example, Miles, 1993). When dominant groups make use of racist stereotypes that are inherently negative, it allows those subordinated groups to be removed from the moral and social space that the racist actor inhabits. In other words, if the 'Other' becomes so far removed from the self, then violent actions towards that group can be condoned. Those who habitually use stereotypes – in effect, excluding the Other from their own moral and social world – can, as Opotow (1990, p. 12) remarks, 'perceive some people as objects and imperceptibly cross a threshold that excludes these Others from their moral universe'. As stereotypes are historical formations, they are not rigid and closed categories but flexible constructions that change according to the specific time and place of their articulation. Further, stereotypes are rarely just negative and are often produced through a complex process that sees the Other in extreme terms: the 'good' Other/the 'bad' Other. Gilman (1985, p. 18) expresses well this 'malleability' of stereotypes when he says:

> Paradigm shifts in our mental representations of the world can and do occur. We can move from fearing to glorifying the Other. We can move from loving to hating. The most negative stereotype

always has an overtly positive counterweight. As any image is shifted, all stereotypes shift. Thus stereotypes are inherently protean rather than rigid.

In the last 20 years, a new theoretical framework has emerged that attempts to give voice to difference and at the same time to reveal the invisible workings of inequality and white (elite, male) power in modern society: Critical Race Theory (CRT). CRT scholars argue that racialization is at work in all areas of Western society, simultaneously hiding the traces of its work and constructing minority groups as belonging beyond the pale of the mainstream. Hylton (2005, 2009) calls for an engagement with CRT by scholars doing research on racism in sport and leisure. Specifically, he asks that researchers meet the challenge of a more critical theorizing and centralizing of 'race' in their analytical frameworks. In the public sector, underlying the development of equal opportunities policies since the 1950s has been a worldview that draws its reasoning from a racialized, race-biased discourse (Gordon, Miller and Rollock, 1990; Nanton, 1989). This discourse has as its basic principle an oversimplified reductionist tenet that reinforces biological arguments, homogeneity and universalism. In sport and leisure policy this marginalization of 'race' is manifest in the lexicon of policy makers who have promulgated a vocabulary that legitimates rather than challenges the notion of 'race', monolithic racial identities and the black 'Other' (Cross and Keith, 1993; Gilroy, 1987; Goldberg, 1993; St Louis, 2004; Thomas and Piccolo, 2000). CRT challenges traditional dominant ideologies around objectivity, meritocracy, colour-blindness, race-neutrality and equal opportunity (Gardiner and Welch, 2001; Henderson, 1998; Nebeker, 1998; Solorzano and Yosso, 2001).

A CRT lens turned upon the mainstream writing of sport and leisure studies throws light upon a domain that traditionally reflects the power and knowledge interests of white social science and enables us to get a clearer understanding of the major structures (power, culture) involved in the organization of sport, which is crucial when racial equality is the ultimate target (Long and Spracklen, 2010). A focus on power processes, white hegemony, racism and equality, accounts for some of the contemporary concerns that have perplexed 'race' theorists and complicated the study of 'race', concerns that at the same time have been consistently ignored by mainstream theorists. For example, the challenge to interrogate phenomena such as whiteness and 'race' in the historical and contemporary developments of sport and leisure, and how processes within sport and society conspire to reinforce or liberate oppressions, is one worth taking (Long and Hylton, 2002; St Louis, 2004). What CRT brings to an understanding of 'race' and racism is the importance of the construction of whiteness as much as the construction of blackness. However, CRT is bedevilled by an over-theorized discourse that has abstracted the problem of 'race' away from empirical analysis of the everyday into the obscurantism of rhetoric. What is needed is a clear theoretical framework, with a clear methodology, that allows us to understand 'race' and the relationship between whiteness and leisure. To get there, we need to consider theories of whiteness in more detail.

Theories of whiteness

Whiteness is a problem for sociologists of the Western world, where the historical circumstances of postcolonialism and globalization have combined to create a number of challenges to the global, social, economic and political power of Western countries. The whiteness of Western countries is taken for granted in historical accounts of each Western country's development. Modern forms of nationalism in each of these Western countries construct imagined communities where belonging is associated with whiteness and nationalism becomes racialized and 'white-washed'. Whiteness is assumed to be the norm, the mainstream, and it is associated with the ruling hegemonic classes, which invented traditions that associated their heritage (their whiteness, their belonging, their usurpation of power) with the natural order of things. For example, in the United Kingdom, the complexity and diversity of the nation was sidelined by the intensely white and elite nature of the guests at the functions of the Queen's Diamond Jubilee. Early scholars of whiteness tried to identify the characteristics that made white, Western countries successful in the first century of modernity: this was the age of Hegelian destinies, of made-up stories about the purity of Germanic heritage and the Social Darwinism of Galton and Hitler (Daynes and Lee, 2008; Kohn, 1995). In this erroneous mythologizing, the white race (or races) was judged to be biologically and culturally pure, the inheritors of some far-off Aryans or Anglo-Saxons who lived lives that white racist ideologues in the nineteenth-century wanted to live. This narrative of whiteness and nation (and masculinity and class, as these social inequalities intersect) continues through the twentieth century and survives today in every Western, white-dominated country: the discourse of forefathers, folklores, blood, Pilgrim Fathers, royal families, Republican virtues, national anthems, national heritage and fatherlands. But this narrative of pure whiteness, pure nationalism, is under threat by alternative histories and discourses of belonging more open to hybridity, exchange and the construction of multiple identities.

A number of scholars and popular authors have attempted to develop theories of whiteness in modernity, or possibly postmodernity (see, for example, Delgado and Stefancic, 1997; Doane and Bonilla-Silva, 2003; Dyer, 1997; Frankenberg, 1997; Garner, 2006; Hage, 1998). Some earlier accounts of whiteness were problematic because they imbued whiteness with essential qualities, suggesting either that there was a white 'race' that was biologically distinct from other races or that whiteness had some culturally essential characteristics that were inheritable (see critique in Dyer, 1997). However, all modern scholars of whiteness acknowledge the contingent nature of whiteness and being white: there is no white biology, no white culture fixed and inherited through family and friends; but there is a whiteness that is taken on and owned by people with power in Western countries, which is used to define political, social and cultural belonging in those countries, and that whiteness, while subject to flux and contestation, is something that is partly imposed and partly learned. Franz Fanon (1967) in his early critique of colonialism wrote of the white mask that some non-white individuals chose (or were forced) to adopt to be accepted in colonial societies. This white mask was of course a psychological construct – it did not mean people became white – but it did mean adopting the particular cultural and social habits of the white imperialists, habits that were universalized as the habits of civilized, respectable society. Fanon's white mask was worn by non-whites seeking to become accepted by white, civilized, imperialist society – but it was a mask that easily slipped, or was easily removed by white elites. We can extend Fanon's account by noting that the mask could also be worn by white people from marginal white groups: in the United States for much of its history, for example, where white immigrant groups have been successfully marginalized over time, but where individuals have moved out of exclusion through changing their names, their habits and educating their children to grow up as White Anglo-Saxon Protestants (WASPs). The mask can be understood in Bourdieusian terms (Bourdieu, 1986): it is the accrual of white, elite cultural capital, which enables the wearer to be taken as a member of the elite white group. For the individual putting on the mask, then, there is much work to be done to keep the mask on: to learn and to accumulate the right capital, to understand the hidden codes, the myths and the secret signs that let them join the secret society. And just as the mask often slips, the mask can also be easily removed by the hegemonic power of elite whites guarding the borders of their symbolic community:

so rich Arabs can buy sports clubs and expensive cars and join polo clubs in New England, but as soon as they are stopped by a police patrol their wealth and cultural capital dissolves away in the stern eyes of the police officer (the agent of white American elites) brought up to believe in the Flag and the nation's destiny.

As discussed in the preceding paragraph, there are a number of white-nesses at work in Western societies. Some white groups are demonized because of their association with the working classes (or the underclass). In the United Kingdom, white, lower working-class people have been labelled as 'chavs', lazy, feckless, fashion-obsessed undesirables who eat a poor diet and who are involved in petty crime. Jones (2011) shows that much of this deliberate marginalization of the white poor uses the racialized, racist stereotypes applied on other occasions to gypsies and the urban, Asian and black communities. He suggests that the intolerant and damning stereotyping of this part of the nation's white commu-nity follows the pattern of capitalist marginalization and control that keeps the rich secure at the expense of the working-class communi-ties impoverished through neo-liberal politics. Such stereotypes of the working-class poor, white or otherwise, are also prevalent in other coun-tries, but with particular historical associations. In France, the political elite and the republican Establishment perpetuate racism and exclusion aimed at (mainly Muslim) people of North African descent (Silverstein, 2004, 2008). In the United States, the complex racial politics of the previous two centuries means derogatory stereotypes circulate about poor whites in the South and West, as well as Latinos and African Americans. In Australia whiteness is associated with 'proper' Australians, the colonizers from Great Britain and Ireland (Hage, 1998), and the legacy of the racial politics of the country in the twentieth century: the marginalization (and massacre) of the Aboriginal population and the more recent racist rhetoric about immigrants and citizens with multiple loyalties.

Steve Garner (2003, 2006; and Clarke and Garner, 2009) has devel-oped a critical sociological critique of whiteness that recognizes the contingencies of history and the multiplicity of identities in the twenty-first century. Garner (2006) begins by situating the growth of whiteness studies in the United States (typified by the work of Frankenberg, 1994, 1997; or Doane and Bonilla-Silva, 2003) in the bipolar racial politics of that country. He then notes the lack of any sophisticated critique of the construction of whiteness in contemporary European sociology, and proceeds to build his own theory of whiteness. He begins by discussing the limits of previous work and the problems inherent in theorizing something that has no ontological status other than a negotiation of power (Garner, 2006, p. 258):

> The writing that does address contemporary matters does so indirectly via the question of labour migration (Miles, 1982, 1993; Noiriel, 1988). This reinstates pre-Second World War white immigration into Europe in a research agenda that had become fixated on black and Asian migratory waves and the hegemonic 'race relations' paradigm of the 1960s and 1970s. The 'new immigration' into Western Europe has further underscored the idea that Others can also be white, or even members of the same nation (Sniderman et al., 2000). Indeed, my departure point is that identities are multiple and contingent (Anthias, 2003) and that racialization in the early 21st century is not fixed by a black–white binary any more than it was in the 16th century (Garner, 2003). Culture is as important as skin colour in racializing discourse, i.e. they are equally valid elements of a 'discursive formation' (Foucault, 1969, 1971). Nominally white Europeans can also be racialized in the process of constructing national identities, as has been the case in Britain with nomadic, Jewish and Irish people, as well as Eastern Europeans. Sivanandan (2001) even coins the term 'xeno-racism' (as distinct from racism sensu stricto, a property of white–black power relations) for this contemporary form of racialization. The whiteness problematic therefore appears to be seeping into the fabric of some of the more challenging work on racism in a European context.

The key authors underpinning Garner's account of contingent and contestable whiteness – what he calls the 'problematic' – are Anthias and Foucault. Anthias (2003) brings to the theoretical framework a critical theory of 'race', identity, hybridity and intersectionality. For Anthias, the psychology of self-identity is constrained by the political and the social: each of us is constantly attempting to make sense of the world around us and our identity in it through recourse to a shifting discursive apparatus – what we are told about ourselves, what others impose on us, what we take from popular culture, what is mediated, what is sold to us, what we are denied. In Western society, identity-making is bound by the historical power relations between different social groups, leading to distributions of power today that favour white over black, male over female, capital over labour, heterosexual over homosexual, able-bodied over people with disabilities. These are the inequalities of power that intersect in the formation of identities and the formation of new inequalities. Foucault (1970, 1972) is used by Garner to support the re-casting of racialization in a wider sense as a discursive formation, something created through discourse, developing an understanding that racialization is a process that constructs and deconstructs identities and power relations through the appropriation of narratives of belonging, community, culture and essentialism. As Garner continues (ibid., p. 264), 'while one dimension of whiteness is its dialectic relationship with non-white othernesses, internal boundaries are equally evident . . . whiteness can best be grasped as a contingent social hierarchy granting differential access to economic and cultural capital, intersecting with, and overlaying, class and ethnicity (Hartigan, 1997a, 1997b, 1999; Jacobson, 1998; Orsi, 1992; Wray and Newitz, 1997), as well as gender and sexuality (Daniels, 1997; Frankenberg, 1994)'. Garner concludes by arguing that whiteness not only is fluid, and related to racialization and intersectionality but is also something that is continually (re)produced and negotiated between groups that might lay claim to the label 'white'.

Garner's theoretical project is an important synthesis of different accounts of whiteness, and much of my own understanding of whiteness is based on his work. This book is a response to the call by Garner for more critical analyses and research about whiteness in contemporary society. However, in aligning his work with Foucault, Garner seems to be allowing a post-structural reading of ontology and epistemology through the process of identity formation. If everything is merely a 'discourse formation' then it becomes difficult to deny the identity-making of some groups that is built on hegemonic power formations: the racism of white people is in danger of being legitimized as just another 'discourse formation' equally valid in the relativism of Foucauldian power relationships. Of course, this is not what Garner believes – but the use of Foucault's 'discourse formation' is problematic if we want to challenge racism, racialization and the hegemony of whiteness. The other problem in using Foucault is the danger of losing sight of the critique of whiteness in the whitewash of Foucauldian (and hence Freudian) jargon.

Conclusions

The politics and popular culture of contemporary Western society show the need for an account of the relationship between racialization, whiteness, hegemony and leisure. The concept of the West itself is problematic. It is an imagined community (Anderson, 1983), something that has no meaning beyond the shared symbols, myths and cultural arte-facts it purportedly covers. Let me qualify that – it does not have any real meaning attached to geographical boundaries or shared experiences, but it has a great deal of meaning in making people think they belong to something that stretches across national borders, something that preserves the idea of white, European, imperialist and capitalist power (Hobsbawm, 1992). The idea of the West owes its origins to the rise of nation-state empire-builders in Western Europe. These nations believed they shared a common culture created out of Enlightenment rationality and Romantic neo-classicism: one that valued art as an expression of national identity; one that encouraged science and industry; one that promoted capitalism and trade; and (less importantly) one that suppos-edly valued a modicum of liberty. Jurgen Habermas (1989[1962]) sees the construction of the public sphere in European civil society as a key stage in the development of Western power and high modernity. It was believed that a particular set of assumptions and ways of thinking were common across the ambitious bourgeois classes of Europe. These were the classes that embraced the idea of progress, of liberalism, of nation-alism and of imperialism, paradoxical ideological standpoints brought together by the successful transition of European societies from feudal to modern. As Edward Said (1978) points out in *Orientalism*, this imagined community of the West was constructed through the stereotyping of its Othered rival, the East (meaning the Muslim world and Asia). Western-ers saw their economic and political supremacy in the Age of Empires as proof of the West's dynamic and positive civilization, contrasted with the weak and degenerate East (Said, 1985). As the political hegemony of the West migrated across the Atlantic to the United States, the idea that old Europe and the New World shared a common culture and a common civil society was taken for granted on both sides of the ocean.

  Despite differences over trade and the politics that transformed into terrible wars, the idea of the West, the cultural and political bond between the United States, Western Europe and the white-dominated former colonies (Australia, Canada, South Africa, New Zealand and arguably the countries of South America) survived into the period fol-lowing the Second World War. Then the growing influence of the Soviet Union in Europe led to a new set of imagined communities: the East being the Communist sphere of influence; the West being the coun-tries that defended democracy, liberalism and capitalism. Both imagined communities laid claim to the heritage of Europe, both used leisure and culture to legitimate their existence and to belittle the achievements of the other imagined community. In their proxy wars in South America,

Asia and Africa, the West and the East acted as white hegemons, extending their own imperial zones and manipulating and oppressing people and entire groups that were perceived to be on the 'Other' side. The East of the Cold War struggled to compete with the West, and political and social upheavals saw Communist governments fall across the world, with the Soviet Union itself splitting up into a wildly capitalist Russia and a number of other successor states. In the first few years following the breakdown of Communism, political theorists smugly anticipated the globalization of Western values and the wholesale embrace of individualism and capitalism. Francis Fukuyama (1992) suggested that the victory of the West would mean the end of any meaningful history (what might be called History in a Hegelian sense). But although the United States extended its hegemonic reach, and Coca-Cola and other American brands such as MacDonald's and Nike are evidence of the success of global capitalism (Klein, 2000), there is still a counter-hegemonic resistance at work in many different places: from the Islamic world, from post-industrial towns and cities in Europe, and from the new economic and political centres of India (Bose and Jalal, 2003) and China (Mitter, 2004). The reaction to the hegemony of the United States and its political allies had led to a recapitulation of the imagined communities. Now, the West is defined as those countries that counter the twin threats of Islamic terrorism, 'rogue states' and Chinese economic power. The West has become a symbol of white hegemony, depending equally on the production lines of Hollywood and the aircraft carriers of the US Navy. The East has again become a place of mystery and strangeness, but it is no longer seen as lazy and decadent – rather, the racialization of the East now plays on white, Western fears of invasion and duplicity.

In this critique of contemporary culture and politics, the importance of leisure as a site of hegemonic, white control and counter-hegemonic, racialized resistance can be seen to be linked to leisure's importance in modern life. The rest of this book is an attempt to demonstrate this, but this analysis of the contemporary West has already pointed out the way in which Western power (white, hegemonic power) rests on the spread of consumer goods and practices of commodified leisure. Hollywood movies provide a narrative of how humans are to behave if they are to become civilized, become Western, to wear the white mask. Whiteness is demonstrated in the drinking of sugary, carbonated drinks and the eating of burgers: the companies might make a big fuss about their commitment to equality and diversity, and people with all sorts of identities might enjoy the drinks and the burgers; but it is still a trick of global capitalism and a diminution of one's freedom to define one's own self-identity against the arrogance of the companies (which presume these white man's drinks and burgers are the best because they sell the most). But leisure offers a route to resist such dominance, too, even if resistance to racialization and hegemonic control only ever seems partial. For example, leisure is an obvious site of diasporic and subaltern resistance. Whether it is the consumption of Bollywood films and local folk music genres, beating former colonial masters at modern sports events or the subversion of tourist resorts through re-claiming beaches and restaurants, leisure offers a multitude of resources for diasporic and subaltern groups. Lashua (2007) shows how First Nation (Aboriginal) young people in Canada have co-opted rap music to explore their own cultures, their identities and their place in modernity. In doing so, they use rap to find a voice and to express their anger at the racism and poverty they face in the street and in wider Canadian society. This counter-hegemonic resistance is the same thing that appears in the agency of the young British Asian women in the work of Aarti Ratna (2010). These young women deliberately choose to play sport not only as a way of proving their own identity but also as a way of challenging prejudice and exclusionary attitudes – in their own communities and in mainstream British society. Leisure becomes a communicative site, an activity that provides individuals with the potential to resist, to play and to politicize that play through radical actions. Leisure is still a form where power structures and discourses have a hegemonic sway, but there is, it seems, enough room around the edge of that instrumental hegemony for people to find a voice and find a space to fight back.

But in resistance and in embodiment of hegemonic structures and ideologies, the problem of racialization often remains. Racialization is as much about embedding and (re)presenting ideas of hegemonic whiteness (Dyer, 1997; Gabriel, 1998) as it is about normalizing racialized Others (Denzin, 2002; Fusco, 2005; Woodward, 2004). Talking about the value of diversity is not the same as having diversity. For as Hage (1998, p. 139) argues, in relation to liberal responses to multiculturalism, 'it is in the opposition between valuing diversity and being diverse that [whiteness] reproduces itself'.

As I will discuss in more detail in the next chapter and later in the book, sociology of sport and leisure has provided a powerful critique of the way in which sport normalizes beliefs about the essence of racial difference. What all sociologists of leisure would like to see in sport is an absence of any racial/ethnic hierarchies, whereas what we have observed amounts only to shifts in those hierarchies: whiteness is still privileged at the expense of other social identities. Until such hierarchies are absent from society, they are unlikely to be absent in sport. At the same time, leisure plays a key role in shaping the 'racial ensemble', which Daynes and Lee (2008) argue is the necessary relationship between racial ideas, racial practice and the belief in 'race'. As they put it (ibid., pp. 138–139):

> What does condition the existence of racial ideas is a process of believing. If there is a belief in race, then there are conditions of validation and reproduction of this belief... Believing can be seen to be a peculiar process; it postulates the existence of an object (the object of belief) for which there is no scientific evidence. Hence believing operates within a specific regime of legitimacy, based upon the pertinence of the process within a set of social relationships.

The whiteness of white people can never be essentialized – there is no such thing as a white race as there is no such thing as a black race (Daynes and Lee, 2008). However, blackness and whiteness, the agency of choosing to identify with one or the other and the instrumentality of defining those who do not belong to one or the other (the Other, as it were) are part of what Daynes and Lee (2008) call the 'racial ensemble', tools used in boundary work, the formation of cultural capital through communicative agency and instrumentalized consumption. Where whiteness differs from blackness is in its link to the dominant side in historical inequalities of power and the useful instrumentality of universalizing white cultural norms as universal norms. In leisure, blackness is inevitably Othered as exotic, and the whiteness of everyday leisure forms is made invisible (Long and Hylton, 2002; Long and Spracklen, 2010). Invisibility is a product of a hegemonic sleight of hand, a trick to fool people into thinking the world is just when it is actually rigged in favour of a small number of elite groups. In Chapter 3, I will discuss studies of 'race' in leisure studies in more detail. In Chapter 4, I will develop a theory of hegemonic whiteness that will expand on the theoretical framework I have formulated from the work of Habermas to understand the meaning and purpose of leisure.

3
Studies of Whiteness in Leisure Studies

Introduction

This chapter provides a review of research in sport, leisure, tourism and popular culture that discusses whiteness. This review probably needs a book of its own to be able to discuss the breadth of literature in any critical depth – so instead of a comprehensive literature review, this chapter will focus only on some emblematic example of the kinds of research that has been done across the broad subject field of leisure studies relevant to the aims of this book. The chapter is arranged into two sections. The first will be a critical discussion of research across the broad remit of leisure studies that explicitly engages with whiteness. Specific research examples from sport, leisure, tourism and popular culture are considered in this section in more detail. The second section of the chapter will critically analyse research from leisure studies that has a focus on blackness, but where whiteness is implied or invoked without any consistent attempt to problematize the concept. I will argue in this chapter that leisure studies, while providing a strong set of research examples that demonstrate the whiteness of leisure, is yet to provide a coherent account of whiteness and leisure.

Whiteness in leisure studies

Sport

Long and Hylton's (2002) paper is an attempt to set out a number of questions, theoretical frameworks and objectives for the study of whiteness in sport. Using Critical Race Theory (CRT), they argue that whiteness in sport, the whiteness of sport's wider culture and networks, is made invisible by the racialized hierarchies of modern Western society. They note that the relative failure of anti-racism policies and campaigns to create more diverse cultures in sports governance is a result of this racialization: whiteness is not problematized and is thus assumed to be the norm within the corridors and committee rooms of sport, while blackness is constantly challenged and questioned and marginalized in its biological place.

Mapping whiteness and sport is the theme of a special issue of *Sociology of Sport Journal* published in 2005. In this issue, the guest editor Mary McDonald (2005, p. 249) articulates a clear understanding of the fluid nature of whiteness and 'race':

> This understanding of the complex racial reordering further impli-cates the performance of whiteness within power relations, especially in demonstrating the capacity of 'whiteness to make virtues of its vices – and vice versa – even as it creates discursive space to decon-struct and demythologize its own socially constructed meanings' (Dyson, 2004, p. 131). Simply stated, this special issue cannot be seen to exist outside of contemporary attempts to restructure a new racial order globally, nationally, and locally, and therefore must also be seen as constitutive of a complex constellation of competing racial projects – supremacist, conservative, liberal, and radical – that also encompass this historical moment, especially in North America and Europe. To engage the complexities, contingencies, and tensions of whiteness in this way is to move away from understandings of whiteness as a 'fixed, locatable identity, ethnicity or even social posi-tioning' to address 'whiteness as a dynamic of cultural production and interrelation' (Ellsworth, 1997, p. 260). That whiteness is shifting and part of a dynamic process is readily apparent when examin-ing the diffuse and changing liberal racial formations and meanings within and among Canada, the U.S., and Mexico, as well as spaces outside of North American borders.

McDonald's justification of the need to investigate the intersections between sport, 'race' and whiteness is proven by research papers pub-lished in the special issue. Douglas (2005) explores the ways in which white people in tennis (fans, administrators and journalists) represent and objectify the black tennis players Venus and Serena Williams. Douglas (2005) argues that such work involves the practice of white-ness as something embodied and lived, not just something ideological: whiteness is constructed through white fans sharing structures of feeling about the sisters and their ambivalent status in the game. Fusco's (2005) contribution to the special issue explores the ways in which whiteness is embedded in sports spaces, specifically the history and architecture of changing rooms. He argues that such spaces are specifically associated with a genealogy of white, heteronormative respectability associated with modernity and a project of purity and hygiene: as such, they become spaces that define normality through white sensibilities, and places of exclusion for marginalized groups. Dennison and Markula's (2005) paper examines the ways in which a press conference for black athlete Haile Gabrselassie became a product of white sport-media hegemony, created to be consumed by white Westerners who wanted to see the athlete fetishized. Cosgrove and Bruce's (2005) paper explores the ways in which whiteness can be seen to be an instrumental force colonizing other identity-formations, through an analysis of the mediatized narratives surrounding the death of the sailor Peter Blake. What is interesting about these four research papers is the dominance of mediatizations of sport in the analysis – the sports-media complex is a key site in the construction of whiteness. The other three papers in the special issue offer more nuanced, critical accounts of the complexity of the construction of whiteness in sport. For Brayton (2005), the whiteness of skateboarding is not simply a product of white male resistance but an appropriation of blackness and a re-construction of racialized inequalities. For Erickson (2005), whiteness seen through climbing subcultures is best understood as an ever-failing signifier that tries to embody wholeness. Finally, King (2005) offers a word of caution about turning to studies of whiteness – that is, it is an academic community already dominated by white academics: such a turn might reify racial boundaries and lose sight of the need to constantly challenge racism and inequity of society, of sport and of sports studies.

Darnell's (2007) paper on the notion of developing social and cultural capital through sports activities shows how such activity constructs an unequal, colonial relationship between the white, Western workers on such projects and the developing nations in which they are working. These projects are seen as ways to improve the lives of people in poorer countries through giving them an opportunity to take part in sports. Darnell uses the work of Foucault to examine the ways in which workers on one of these projects see themselves as benevolent in-comers, spreading the gospel of good character and fair play through taking part in sports activities. For the workers, their mission is to bring civilization to the primitives living in the dark heart of the 'Third World': the workers are white Westerners taking their sports to the foreign Others. For Darnell, whiteness is something that is associated with hegemonic power, cultural capital and the historical inequalities between the West and the rest of the world. The historical nature of colonialism and the post-colonial settlement adds to the analysis as sports development becomes a substitute for real, radical change. As Darnell (2007, p. 514) concludes:

> Through sport and development, Whiteness is (re)confirmed as an intelligible and recognizable subject position, one characterized as benevolent, rational and expert. This position is intelligible in opposition to bodies of colour, recognized as marginalized and unsophisticated, yet simultaneously and continuously grateful for the boons of development.

Leisure

The literature on whiteness and leisure is limited – and much of it is actually about the whiteness of leisure researchers who are researching 'race'. Watson and Scraton (2001) explore their own whiteness and the obstacles they face in their interaction with South Asian women in research undertaken to study their leisure lives. They are concerned with the ways in which they might make assumptions about the role of leisure that come from their white education and white, middle-class British cultural space. They are also concerned that they are seen by their South Asian respondents as being different because of their whiteness and unable to empathize and understand their experiences as working-class South Asian women. The obstacles faced by white leisure researchers reappears in my own work on rugby league (Spracklen, Timmins and Long, 2010). Rugby league is a very white (and male) northern English sport, played mainly by white people and watched by working-class white people. In the paper, we discuss the problems of researching racism and racialization as individuals with racialized identities: the whiteness of two of us and the blackness of the third author. The whiteness allowed the two white researchers in our team access into rugby league's imagined, imaginary community; but it restricted their ability to gain the trust of black respondents who had suffered racism. For the black researcher in our team, his blackness immediately marked him out as an Other (even though ironically he was the only one of us to have played and coached the game at a senior level).

Long and Hylton's (2002) paper – cited above – marks out an attempt to begin to problematize the construction of whiteness in leisure (and in leisure studies). However, whiteness in leisure is generally under-researched and under-theorized. There are numerous papers that mention the whiteness of leisure participants, but hardly do any of these papers attempt to think systematically and critically about the racialization of leisure and the normalization of whiteness. The most salient contributions to theorizing whiteness in leisure studies come from the *Journal of Leisure Research*, which published a challenging keynote address by Mary McDonald (2009) that led to a special issue of the journal in 2009 on whiteness, difference, racism and leisure. McDonald's paper surveys the literature on whiteness and the historical debates about the nature of racialization and then discusses the few examples of leisure research that had problematized whiteness (extending her survey freely to sport and tourism). In her discussion she suggests that leisure researchers have started to address the problem of whiteness (and its invisibility and hegemony) and the way leisure intersects with such identity formations, but she is cautious about the way forward for leisure studies (McDonald, 2009, p. 17):

> Whiteness works through (among many others) contemporary discourses of race, public policy, national imaginings, changing local–global relations, media representations, and shifting legal statutes. Attempts at normalization and linkages with intersections of gender, class, sexuality and nation additionally expose whiteness as contradictory and contingent with respect to time and space. Given these insights, the diverse meanings and forms of leisure – parks and recreation, sport, popular culture, arts, tourism, etc. – all offer important sites to reveal 'whiteness as unfrozen, whiteness as ensembles of local phenomena complexly embedded in socioeconomic, sociocultural and psychic interrelations' (Frankenberg, 1997). Within this conceptualization and via leisure, 'whiteness emerges as a process, not a 'thing,' as plural rather than singular in nature' (Frankenberg, 1997)... And while there certainly is much work to be done to carefully analyze the various ways whiteness is asserted and resisted via leisure practices and contexts, caution should be exercised as well. Indeed Winnubst (2006, p. 9) has characterized similar analyses as 'dangerous' given the powerful universalizing and normalizing tendencies of whiteness. This suggests analyses might produce several unintended consequences 'from playing into cultural discourses of white supremacy, to uncritically fixing white superiority, to reinscribing whiteness at the center of concern and focus'.
>
> (Winnubst, 2006, p. 9)

In the special issue that follows, Arai and Kivel's (2009) editorial suggests that the interest in whiteness, difference, racism and leisure evidenced by the number of submissions they received is some evidence of leisure scholars becoming interested in examining the key meaning and purpose of leisure in racialization. The papers in the special issue do advance the debate about whiteness and leisure. Roberts (2009) responds to McDonald (2009) by supporting most of her key arguments and suggestions, but builds an alternative view of the purpose of leisure research based around the importance of fighting for justice against injustice. Mowatt (2009) warns of the danger of failing to challenge assumptions about whiteness, the taken-for-granted status of whiteness in Western society and the ways in which whiteness is made unproblematic in leisure studies. Three research papers explore the racialization of leisure spaces and leisure forms, looking at the construction of whiteness and other racialized identities (Erickson, Johnson and Kivel, 2009; Richmond and Johnson, 2009; Yuen and Pedlar, 2009). Kivel, Johnson and Scraton (2009) discuss the intersectionality of racism and other forms of prejudice and argue that leisure is a site of hegemonic discourses and 'Othering' – while warning leisure scholars of falling into the trap of essentializing contingent and contested identities. Arai and Kivel (2009) claim that this interest is evidence of a new, fourth wave of research into racialization and whiteness in leisure. As they argue (ibid,. p. 464):

> The fourth wave of race research may be framed as including an analysis of the social construction and deconstruction of racial categories. The fourth wave places emphasis on re-examining race and racism, rather than just ethnicity and cultural differences. This wave contextualizes discussions of race and racism within theoretical frameworks which enable broader discussion of social and structural inequalities, power, ideology and white hegemony. These shifts also raise the call for the use of more diverse methodologies for examining race and racism in leisure studies.

Whiteness in leisure and leisure studies, then, is beginning to explore and problematize whiteness (I would not be writing this book if it were otherwise), but as yet there is little criticality to the problematization of whiteness – or the multiplicity of whitenesses. Some scholars advocate a CRT approach to understanding whiteness, some follow post-structural trends and others remain rooted in Marxist approaches (Long and Hylton, 2002).

Tourism

The whiteness of the tourism industry is not mentioned as often as might be expected. Cara Aitchison (2001) critiques the industry for its post-colonial Western whiteness. She argues that brochures and other marketing material sell a myth of the white, Western traveller setting out and encountering exotic, dark-skinned Others. Drawing on post-structuralist feminist theory, she suggests whiteness in tourism is something that is constantly re-made by the industry and by tourists, and that making is then made invisible, so that the practice of whiteness and Othering becomes normalized. This interplay of whiteness being created through the Othering of non-white people – locals, strangers, hotel staff and so on – is evident in the work of some other tourist studies scholars. Tim Edensor (2002, 2004, 2006) has made substantial contributions to understanding the shifting sands of nationalism and social identity in the tourism and travel public sphere. He asserts that the tourist and the traveller perform dominant, hegemonic roles: the American abroad, for example, throwing their credit cards around; or the gap-year Australian living with other Australians in a shared flat in Bangkok. In all these situations, there are complexities of power and identity-making between tourist and local, but the flow of exchange is almost entirely one way: tourism becomes a way of reifying whiteness and Western nationality by shaping what are fluid and dynamic local identities into exotic, Othered, static structures (Edensor, 2006). In his paper exploring the invisibility of whiteness in the English seaside resort, Dan Burdsey (2011, pp. 540–541) writes:

> Racialized meanings and demarcations of Otherness are mapped and inscribed onto particular seaside spaces which, through their use and (re)imagination, interpellate particular bodies as (in)organic members, and facilitate claims as to who belongs and who does not (Alexander and Knowles, 2005; Puwar, 2005; Sibley, 1995). Taking this into account, the minority ethnic resident or visitor at the English seaside might therefore be conceptualized as a 'stranger'. As Ahmed (2000) argues, the stranger is not someone who is yet to be encountered. Instead, the fact that they have been recognized as a stranger in the first place, as opposed to going unrecognized, is premised on the reality that they must have already been faced in this setting. In other words, the stranger has been 'here' before, and because of previous engagements and the manner in which this space is delineated and controlled, they are 'already recognised as not belonging, as being out of place' (Ahmed, 2000: 21). For this to

occur, the seaside must be continually demarcated as a place where those who are 'not strange', i.e. white British communities, *do* belong. These relationships are enacted within the spaces, built environment and leisure facilities of the seaside, as 'encounters between embodied others… involve *spatial negotiations* with those who are already recognised as either familiar or strange'.

(Ahmed, 2000: 24)

Burdsey's paper suggests that tourism spaces and places remain white only through a mixture of Othering and identity-formation among the majority white visitors to those spaces. In domestic tourist spaces in the West, whiteness is also something that emerges from the encounter between visitors and mainly white locals. On the beaches of England, Burdsey finds white British families at ease with the rituals of holiday-making and the peculiarities of the seaside. They are in turn surrounded by white people who have chosen to retire to the seaside resort because it is less multicultural, and by the locals who see themselves as white people who easily identify with white visitors.

Melissa Hargrove (2009) draws on Bourdieu (1986) to help understand the interaction between whiteness, white racism, Othering and the shaping of heritage narratives in her work on black communities in the South of the United States. She suggests that a social field of whiteness operates in discourses of urban regeneration that limit the agency of local, black communities. One of the ways in which black communities are constrained and made to serve the needs of the social field of whiteness is heritage tourism. This white control of heritage tourism, says Hargrove, is a form of white racism that promotes a particular, white-washed account of the past that objectifies the local black communities and ignores the racism and prejudice that shape their lives in the present day. As she puts it (Hargrove, 2009, pp. 101–102):

> The political struggle over representation within Charleston's heritage tourism industry does not represent an anomalous predicament of racist historical revision. On the contrary, cities such as Charleston are among a growing number of spaces representing this delusional creation termed the 'New South' (see Gaston 2002). In the three decades following the Civil War, the habitus of racism gave birth to this mythic ideology capable of valorizing the southern past (Gaston 2002). Within this New South of the White imagination, plantations are remembered as glorious farms showcasing the entrepreneurial expertise of the White planter class. Slaves are recalled as dangerous,

infantile, and subhuman; thus in desperate need of White paternalistic guidance. Slavery, as an institution, was a necessary evil.
And, most importantly, the Civil War was fought in defense of freedom and state's rights, bearing no relation to the perpetuation of
White supremacy... In interviews from Georgetown, South Carolina,
to Fernandina, Florida, Gullah/Geechees expressed frustration regard-
ing these elaborate fabrications of the 'Old South,' concocted to feed
White racist fantasies of antebellum life and leisure. These reinvented
representations of history, most suggest, have but one golden rule: do
not offend the White visitor! White paternalistic control over the heritage tourism industries, therefore, guarantees the necessary 'white
point of entry' (see Page 1999) demanded by tourists.

What can be seen in the literature on tourism is an awareness of the
unequal encounter between locals and travellers and the spaces in which
those encounters take place – and the purpose of the more commodified end of the industry as sites of residual, hegemonic white, Western
culture (Edensor, 2006). In the book chapter by Rodriguez (2001), the
success of heritage tourism of the American West – the cowboys and the
frontier towns, recreated in the present day – is a symbol of the fear of
white Americans about their future status in the United States. White
Americans visit the American West to see a world that is populated
by white people fighting and defeating Native Americans ('Indians'),
a world where white people are superior and successful in carving
out spaces that are free from troubles for themselves. White American
tourists can be sure of their whiteness and their ownership of the country through constructing myths of cowboys and frontier life. Of course,
this account of whiteness and tourism suggests that tourism has become
a place where white people can retreat from the multicultural reality of
the present, where white hegemonic power is being challenged. Other
forms of tourism and travel suggest that hegemonic power is still held by
white people. O'Connell Davidson (1996) has explored the nature and
extent of sex tourism in the Caribbean and argues that such tourism is
the product of hegemonic forms of whiteness and Western hegemonic
power: white people travelling to an exotic post-colonial space to exploit
black sex workers. While a dramatic form of tourism, sex tourism shapes
whiteness and Otherness through hegemonic encounters.

Popular culture

There have been dozens of research papers, books and book chapters
on whiteness in popular culture (see, for example, Clarke and Garner,

2009; Kellner, 1995). In most of these studies the power of whiteness is contested and challenged by counter-narratives in popular culture, but ultimately popular culture is viewed as a commodity in which white power is retained: making racialization invisible and normalizing white interests as the interests of wider society. Frankenberg's (1997) edited collection is a sound analysis of popular culture's key role in the (re)production of a plurality of whitenesses. Fishkin (1995) provides a useful history of debates about whiteness and popular culture in the United States and demonstrates that whiteness is always contested, contingent and constructed from myths and Othering. As he explains, debates about who is white, and how Othering is used to make whiteness, are intertwined with popular cultural forms (Fishkin, 1995, p. 436):

> Both 'whiteness' as a social and political construct in Roediger's terms, and the kind of complex cultural exchange to which Lott paid such close attention, figure in the work of scholars from the humanities and the social sciences in the early 1990s who, in a variety of ways, asked the question, 'are Jews "white"?' Sander Gilman observes, for example, in his 1991 book The Jew's Body, 'For the eighteenth- and nineteenth-century scientist, "the blackness" of the Jews' was assumed. Gilman notes that the author of an 1850 tract that became 'one of the most widely cited and republished studies on race,' for example, referred to 'the African character of the Jew, his muzzle-shaped mouth and face.' Because of skin color and facial characteristics, Jews were 'quite literally seen as black.' In his final chapter, Gilman examines the genesis of the 1927 film 'The Jazz Singer' and the image of Al Jolson, playing a cantor's son, at the film's close 'on bended knee, [singing] "Mammy" in black-face for his hugely successful Broadway opening.' Gilman refers to 'the long vaudeville tradition of white performers putting on black-face' and then returns to his original question: 'Are Jews white? Or do they become white when they...acculturate into American society, so identifying with the ideals of 'American' life, with all its evocation of race, that they at least in their own mind's eye become white? Does black-face make everyone who puts it on white?' In his 1992 article, 'Blackface, White Noise: The Jewish Jazz Singer Finds His Voice,' Michael Rogin further explores the complex matrix involving whiteness, blackness, and Jewishness in 'The Jazz Singer,' a film which, Rogin observes, 'appropriated an imaginary blackness to Americanize the immigrant son.' And in her innovative 1994 essay entitled 'How Did Jews Become

White Folks?' anthropologist Karen Brodkin Sacks explores the economic, social, and psychological impact of post-World War II changes
in real estate practices on Jewish and black Americans.

The intersection between whiteness, popular culture and leisure is rarely
explored, but there are some excellent attempts to begin the critical
analysis. Kusz (2001) interrogates popular culture and media representations of sport to make the claim that similar narratives of 'crisis' appear
about the loss of status and power of young white American men. This,
he says, is a product of a particular dominant white discourse that
emerged in the 1990s, when white people started to fear the loss of
power and hegemony in the shifting sands of the country's politics. He
begins his paper by referring to the pop-pink band Green Day, who had
hit the American charts with their song 'Minority', in which these white
men sing of being victimized and marginalized. For Kusz, the song is a
reflection of a wider trend in 1990s American popular culture: the rise of
young white men singing songs about how they are marginalized and
ignored. This trend, says Kusz, is replicated in the other areas of popular culture which he analyses. First, in polemical books questioning the
decline of American youth and the problem of school shootings, Kusz
finds a narrative that identifies the victims as young white men. Second, in analysing debates in a sports magazine about the lack of white
athletes, he again finds this worry about absent or marginalized white
victims, young white men who have lost their 'natural' place in front of
the pack. The crisis of young white masculinity is, of course, a narrative
device. White people retain power in the United States and control its
popular culture industry, so for Kusz this discourse is a double-bluff, an
argument that makes claims about marginalization when such displacement of power has not taken place. There are many disenfranchised
young white men in modernity but they are in such a position because
of neo-liberal economic and global capitalism.

Avila (2006) is a social and cultural history of the formation of white,
middle-class suburbia in Los Angeles, a type of fantasy land that was
shaped by 'white flight': moves by white families away from the non-
white communities of the inner city. The shift to suburbia was bolstered
by the construction of the white suburb in the popular imagination,
through the use of such settings in films, television and literature in
the second half of the twentieth century. White families wanted to be
in places where only other white people lived, homes with neat green
lawns and white picket fences, with driveways and garages for their cars.
For Avila, the whiteness of such spaces was defined by the inequalities

of capitalism, which through the period fostered racism, prejudice and exclusion. White picket fences became a popular cultural symbol for exclusivity and white aspirations. And suburban life became a symbol of white American cultural hegemony.

Blackness in leisure studies

Sport

The (re)production of blackness in sport has been the subject of a huge range of books, special issues of journals, journal articles, book chapters and conference proceedings. Some researchers have concentrated on the nexus between racialization and inequality, on racism and exclusion; others have explored the ways in which blackness is achieved and constrained by the tension between hegemonic powers and individual agency. In the edited collection *'Race', Sport and British Society* (Carrington and McDonald, 2001a), for example, there are research chapters that focus on blackness and racism in the sports of football/ soccer (Back, Crabbe and Solomos, 2001; Dimeo and Finn, 2001), cricket (Carrington and McDonald, 2001b) and rugby league (Spracklen, 2001). These four chapters capture professional team sports with fans and participants in a range of settings, all of which frame blackness as something alien and Othered to the 'mainstream' of the sports. Another three chapters in the book concentrate on the ways in which blackness is essentialized through various mediatized discourses: through the false science of racial biology in sport (Fleming, 2001), through the conflation of nation and 'race' (Marqusee, 2001) and through ethical and ontological problems with the notion of 'colour-blind' laws that do not adequately address issues of racism (Gardiner and Welch, 2001). The final chapters of the collection set out to challenge discourses and contest identities, to greater or lesser effect: Johal (2001) writes about the passion for football among South Asians in the United Kingdom, which is nonetheless marginalized by the hegemonic sports culture; Scraton (2001) discusses the intersections between 'race' and gender; Lindsey (2001) explores her own experiences as a black, female sports journalist; and Searle (2001) reflects on the impact and importance of the work of CLR James.

  Ben Carrington has been one of the most insightful contributors to the sociological understanding of sport, racism and blackness. In Carrington (1998) he explores the way in which black masculinities are constructed through the involvement of black men in sports. In his work on cricket in the United Kingdom (Carrington and McDonald,

2001b) he has shown how black British communities with diasporic roots in the Caribbean have used the sport to maintain a sense of social identity in the United Kingdom and a safe place where they do not have to suffer racism or perform the carefully defined roles demanded by white British society. In his work on the sports-media complex in the United States and the United Kingdom, Carrington (2010a) explores the way in which bodies and individual athletes are racialized, marked for their difference and kept in their place. In 2004, Carrington was a guest editor of a special issue on 'race' and sport of the journal *Leisure Studies*. In this special issue, there are original research papers by Woodward (2004) on racialization and masculinity in boxing; by Garland (2004), King (2004), Farred (2004) and Crabbe (2004) on different aspects of football; and a critique of racial science by St Louis (2004). In his introduction to the special issue, Carrington (2004a, p. 3) highlights the complexities of sport and racialization in this century by arguing:

> Sociologically we need to avoid the twin dangers of the sentimental conservatives, who believe that the mere fact of white fans singing the names of black athletes demonstrates the end of racism, as much as we do against those pessimistic commentators who dismiss any shifts within the realm of the popular as merely ideological.

In his later work, Carrington (2010b) has argued that the inclusion of more black participants in sport in the West, the successes of individual black athletes and the wider political struggles of African Americans and other black minority ethnic groups have not stopped sport from being the site of a political and cultural battle between black people and the still-dominant white elites. Carrington's account of racism in sport and the power struggle over definitions of identity is firmly situated in a Marxist critical lens (Carrington and McDonald, 2008), refracted (as discussed earlier) through the concern over the relationship between culture, structure and power seen in the work of Stuart Hall (1993).

Kevin Hylton's research project on sport, 'race' and racism (2005, 2009) is situated in CRT (see Chapter 2). Hylton sees the politics of sport and 'race' as the contestation of the spaces, policies and practices of sport by different social groups. In the West, sport is a vehicle for the maintenance of whiteness: hence, historically, blackness has been marginalized or reduced to an exotic Other, defined forever by colour and heredity and physicality. However, this concern with maintaining whiteness through making the process of whiteness-making invisible is not a necessary feature of sport, only an accidental one. For Hylton, sport has the potential to be more inclusive, more welcoming, more fluid in its power relationships and more open to the possibility of negotiations of other identities (Hylton, 2009). As such, Hylton sees sport as a potential leisure activity in which black and other minority ethnic group identities can be created, celebrated, supported and reproduced: freed from the hegemonic power of whiteness and heteronormativity, sport can be the crucible for the formation of alternative identities and communities, ones that are hybrid, fluid but meaningful for those who want to use sport in this socializing way. For this to succeed, Hylton argues that sport needs to recognize and understand difference and show that it is committed to equality and tackling racism through positive action and the radical re-distribution of power within the structures and cultures of sport (Hylton, 2005; Long and Hylton, 2002). Identity in sport then becomes a matter of personal agency and lived experience rather than domination and constraint, and blackness can be constructed in a positive way that recognizes historical roots and (post)modern identity-formations. This potentiality for sport as a site for the positive construction of minority ethnic identity and community is recognized in the work of authors researching British Asians and sport, such as Scott Fleming and Aarti Ratna. For Fleming (1991, 1994, 1995, 2001), sport is an activity that historically perpetuates racism, discrimination and exclusion – but it is also an activity that gives meaning and purpose to the lives of the young South Asian men he researched in his book *Home and Away* (Fleming, 1995). The construction of their South Asian (British Asian) identities is partly articulated through their involvement in sports they play among themselves. These men also use sport to construct their gender. This is also the focus of the work of Ratna (2010): she has explored how sports such as football and cricket have given British Asian women the opportunity to challenge dominant discourses about their passivity and invisibility in the public sphere. The women in Ratna's research have faced racism and prejudice through taking part in sport (from white people and sometimes from their own communities), but their sports participation has seen them empowered to shape their own identities, their own narratives of belonging and their own futures.

Leisure

Blackness in leisure has been the subject of a wide range of research papers and theoretical essays and books – many of which have been mentioned in previous chapters and in the preceding sub-section on sport. In the special issue of *Journal of Leisure Research* on whiteness from

2009 there are three papers that explore different aspects of non-white racialization: the construction of whiteness and blackness in visiting the Rocky Mountains (Erickson et al., 2009); the experiences of ceremony for Aboriginal women in prison (Yuen and Pedlar, 2009); and the racialization of leisure in an American prison (Richmond and Johnson, 2009). These papers built on an existing body of academic literature on blackness and leisure. Philipp (1995) strongly critiques existing models of leisure for failing to take account of racialization and sets out to challenge the notion that leisure in the West is something that is unracialized: for Philipp, it is the task of leisure scholars to embrace both racism and 'race' to better understand the marginalization of black people in leisure settings. This strong critique of leisure research, especially the social psychology of leisure, is picked up by Floyd (1998), who shows how the dynamic natures of 'race' and ethnicity have been simplified into unproblematic groups in much of the research (which in turn has failed to explore the power inequalities at work in leisure that shape differential patterns of engagement). Philipp's call for changes to leisure research is taken up by Shinew, Floyd and Parry (2004), who attempt to map the different leisure experiences of black and white Americans. Their work, however, is problematic in the way that blackness and racialization are something described rather than explained.

Henderson and Ainsworth (2001) reflect on the challenge of researching the leisure lives of people from minority ethnic groups, and in doing so demonstrate how important leisure is in the construction and maintenance of black identities. For the African American and American Indian women they research, having access to leisure time and spaces is a key issue that constrains their ability to find value in leisure, but leisure remains fundamental to their meaning-making agency: leisure, while a site of racism and exclusions, serves as an activity that the women feel they own and which gives them a sense of worth and a sense of place. That said, there are challenges to establishing the extent of racism faced by the women in the study, as Henderson and Ainsworth (2001, p. 29) conclude:

> In the CAPS study, we found that discrimination and prejudice may manifest themselves in subtle ways rather than being overtly discussed. Discrimination was not often articulated when direct questions were asked, but examples sometimes emerged during the interview. We cannot assume that this lack of explicitness means that discrimination did not exist in the lives of these women and that race did not matter.

Hylton (2005) echoes Phillip and criticizes leisure scholars for their failure to account for racism and actively campaign for the inclusion of others and the re-shaping of racial hierarchies and white hegemonies: blackness is as equally problematic and contingent as whiteness, and as such, says Hylton, the methods of CRT need to be applied at all times to understand the ontological and epistemological problem of leisure. Swinney and Horne (2005) undertake an analysis of the ways in which local authorities in Scotland have revised their leisure policies to ensure they serve their ethnically diverse populations. In this paper, the authors note the importance of leisure and the failures of policymakers to take into account the complex needs of minority ethnic groups.

Tourism

For minority ethnic people in the West, tourism preferences are constrained by fear of racism, actual racism and racial prejudice within the white-dominated tourist industry: there is evidence that there are differences in tourism preferences between hegemonically dominant white people and black people in the United States (Philipp, 1994). Gibson (1994) has shown how the boom industry of active sport tourism is still highly stratified and an activity for elite whites against poor whites and minority ethnic people. However, there is research on minority ethnic tourism that suggests there are new dynamics and encounters that offer minority ethnic groups a chance to enjoy a break and play with their own histories and identities. Klemm (2002) has surveyed the holiday habits of British Asian people in the city of Bradford and concluded that their habits are very similar to the British population as a whole, but they are still ignored or overlooked by the tourist industry. Grant (2005) explores how the city of Philadelphia in the United States has become interested in actively attracting minority ethnic tourists. Finally, Pierre (2009) shows how African Americans and others from the black diaspora are holidaying in Ghana to explore heritage tourism sites associated with the slave trade. Tourism and travel can be positive activities for the construction and maintenance of alternative identities, and can provide a space where racism and prejudice are absent or challenged. However, tourism is still an industry that serves the needs of white Westerners who wish to travel to a place where they can relax – either a domestic resort where they can mingle with other residual whites, as Burdsey (2011) shows, or a foreign resort, where they can become the imperial white colonist through the objectification of the local population.

In the paper discussed earlier in this chapter, Aitchison explores the racialization of tourism encounters and the construction of blackness as Otherness. After exploring a number of examples of Othering in the industry's marketing literature, and explaining that this marketing creates fantasies of exotic foreign spaces and people in the minds of white, Western, bourgeois consumers of overseas tourism, she provides a critical discussion of how Othering works (Aitchison, 2001, p. 136):

> Rose (1996) illustrated that the binary distinctions between Self and Other, and real and imagined space, are part of what Butler (1990: 13) has referred to as 'the epistemological, ontological and logical structures of a masculinist signifying economy'. Thus, the construction of dualisms or binary opposites is inherently related to the construction of the Other. Developing on work by Cixous (1983), it is possible to identify three fundamental relationships within this process. First, the construction of the Other is dependent upon a simultaneous construction of 'the Same', or something from which to be Other. Second, this relationship is one of power whereby that which is defined as 'Same' is accorded greater power and status than that which is defined as Other. Third, that which is defined as Other is attributed a gender and this gender is usually feminized: 'the concept of women as Other ... involves the central claim that Otherness is projected on to women by, and in the interests of, men, such that we are constructed as inferior and abnormal'.
>
> (Wilkinson and Kitzinger, 1996, pp. 3–4)

This account of Othering is drawn from post-structural feminism but is applied by Aitchison to explain the Othering of non-white individuals in tourist encounters and the industry's marketing. The inequality of power entails the construction of static, inferior stereotyped, biological Other races and nationalities. The reality of local people's fluid and contested identities becomes fixed by the white gaze. Aitchison continues (2001, p. 137):

> Tourist destinations as sites for tourists, and the people within them as sights for tourists, are frequently rendered Other by a tourist industry that has developed an unsigned colonialist and gendered hegemony in the form of a set of descriptors for constructing and representing 'Tropical Paradise'. These descriptors signify a colonial legacy where places are viewed as mystical or treasured landscapes preserved by time to be explored, and often exploited, in their natural

state. The people within these landscapes are frequently portrayed as passive but grateful recipients of white explorers from urbanized and industrialized countries searching for their authentic origins.

There is a wealth of literature in tourism studies that explores the racialization of the tourist encounter and the construction of stereo-typed Others. A few examples are sufficient to understand the literature. Munt (1994) examines the rise of green/sustainable tourism and in particular the development of eco-tourism, in which rich Westerners travel to developing countries to do good work saving forests and animals. This eco-tourism is critiqued by Munt as a colonialist encounter, where mainly white people from the West act as missionaries and imperial officers, travelling to poor countries where the local non-white people are crudely stereotyped as ignorant savages in need of Western aid. The white eco-tourists see themselves as doing good but see their role as bringing Western enlightenment to places they think of as dark and backward, populated by lazy blacks. Eco-tourism, then, is a powerful force for the construction of Othered notions of blackness, stuck in its ways and bound by nature, saved by the intervention of the hegemonic white imperialists. Gotham (2002) shows how blackness is fetishized in the commodification of the festival of Mardi Gras in New Orleans: the city has become a site where racialization is hardwired into the images used to sell the city as a bacchanalian space of physicality and voodoo. Finally, Rivers-Moore (2007) critiques the way promotional campaigns by Costa Rica aimed at foreign tourists use problematic images and symbols of exotic blackness. What these pieces of research have in common is an awareness of the inequality of the tourist encounter: all tourism is predicated in an unequal relationship, where those with money and cultural capital exchange it for an experience in a place where people have less money; and nearly all tourism is associated with the movement of rich, white, Westerners into other spaces that are routinely stereotyped in some way or another so that the tourist sees what they want to see.

Popular culture

Blackness and racialization in popular culture have been the subject of an enormous range of literature. Hebdige (1979) explores the way musical genres construct racialized group identities, both for those in the subcultures and those outside the group. Lipsitz (1990) explores the construction of national and racial identities through the consumption of popular music and radio programmes. Hall (1993) identifies popular culture as the key site for racialization and the construction of belonging and exclusion. Young (1995) discusses the post-colonial imagination in popular culture and the struggle for and against hybridity in popular culture. Gilroy's (2000) book, mentioned in the previous chapter, is a comprehensive critique of the construction of racial identities and the racialization of popular culture. Edensor (2002) extends the analysis to national identity as well as racial identity, arguing that all popular culture is a form of shaping identity and normalizing the ruling hegemonic power.

In the exploration of racialization, popular culture and leisure, there are a few examples in the research literature. Bruce (2004) shows how commentators on television sports programmes routinely construct racialized Others in the narratives they present, contributing to popular cultural myths about black physicality. As she concludes (Bruce, 2004, p. 875):

> Live sports television is a high-pressure, high-stakes environment, particularly for commentators who are the public faces and voices of the production team. It should not be surprising that, under pressure and on a live stage, commentators draw upon widely circulating racial ideologies. Indeed, how could they not, given their location in a profoundly racialized culture? Racist ideologies may be more likely to appear when commentators are under stress or thrown into unfamiliar situations without their usual depth of knowledge about participants. In these situations, perhaps more than in others, the commentators are 'spoken by ideology'. And it is the ideologies of socially dominant groups that are most likely to be produced or reproduced in these heat-of-the-moment slippages. For, as Rowe points out, at these moments 'we are more likely to find out what...Whites think about Blacks than vice versa'.
> (1999, citing Blain and Boyle, 1998: 371)

This is a profoundly negative view of how popular culture reproduces racism and prejudice, and racial stereotypes. But popular culture and leisure offer potential sites for resisting such dominant discourses. Green and Singleton (2006) interview women from a British Asian background, as well as white British women, about the constraints on their use of various leisure spaces. In their analysis of the responses, Green and Singleton show that leisure spaces have similar problems about risk for both groups of women. They also show that the Asian women use leisure and the consumption of popular culture to define their hybrid British Asian identities. Finally, Lashua (2007) examines the value of rap music in the leisure lives of young Aboriginal students in Canada. He tells their story by allowing the artists themselves to express themselves, creating what he calls a performance text – lyrics and narratives about their everyday lives. This paper shows how this form of music empowers the young students to condemn racism and exclusion, but also to be proud of their multiple identities.

Conclusions

This brief overview of some of the key theories and research on whiteness and leisure demonstrates both a plurality and absence. There is an abundance of work on racialization, whiteness, blackness and leisure, sport, tourism and popular culture. This review touches the surface of a huge field of literature – there are special issues of journals, books and even the journal *Whiteness* that I have excluded from the discussion but which could have been referred to in a longer analysis. This plurality is a strong point of the literature: there is a huge amount of research being done that sheds light on particular forms of leisure, specific sports or cultural practices. But the plurality, somewhat paradoxically, demonstrates an absence. Firstly, there is a paucity of research that attempts to synthesize arguments and theoretical frameworks from across leisure studies. Secondly, there is much more work needed to build a coherent picture of the construction of whiteness in leisure: more attention needs to be given to the different ways in which leisure is constructed. Thirdly, there is a need to build a coherent theory to account for the mechanisms at work in leisure. How is whiteness reproduced? What does that mean? How many whitenesesses are being created – and which ones count? Who or what gets to decide what form of identities are favoured (made normal) and marginalized (made Other) in leisure? How much agency do people have to use leisure to resist hegemony, commodification and racialization? Leisure studies, while providing a strong set of research examples that demonstrate the whiteness of leisure, is yet to provide a coherent theoretically framed account of whiteness and leisure. The next chapter will provide a theoretical framework to understand the meaning and purpose of leisure in the construction of whiteness.

4
A New Theory of Whiteness

This chapter provides a new theory of whiteness based on Habermas' insights into communicative reason and instrumentality. I will first discuss Habermas' contribution to our understanding of social identity and power, positioning his work in wider critical theory. I will then discuss criticisms of Habermas from the post-structuralist school, which position Habermas' defence of the Enlightenment project as a retreat into Eurocentric discourses of white, male power. I will show that such criticisms are philosophically naïve and suffer from a self-contradiction in their argument. Rather than reifying white privilege, Habermas' defence of communicative reason provides a space in which such privilege can be challenged – when such privilege becomes identified with a form of instrumental rationality about 'race'. Whiteness becomes an all-pervasive instrumentality, which, like capitalism, threatens to consume the entire world. The existence and survival in the Academy of counter-narratives of 'race', predicated on communicative rationality, shows that the Enlightenment – while flawed in history – remains a durable ideal of free inquiry.

Habermas and critical theory

I have already outlined the development of Habermas' thinking in my previous work for Palgrave Macmillan, and in a number of published research papers (see, for example, Spracklen, 2009, 2011; Spracklen and Spracklen, 2012), so I will not spend too much time repeating myself. However, it is probably worthwhile at this stage returning to that discussion to outline the key stages and ideas.

Habermas has been an enormous influence on the development and direction of European Critical Theory. He is an inter-disciplinary scholar, sensitive to history, framing his ideas in the language of philosophy and cultural studies, contributing to our understanding of sociological and political issues such as globalization and postmodernity. He began his career in the shadow of Adorno and Horkheimer at the Frankfurt School. They taught Habermas about the instrumentality of modern, popular culture and the slow commodification of the world. They also taught him about the meaning and purpose of proper high culture, its moral value as a site for the exploration and construction of humanity – Adorno, in particular, wished to defend classical music and European art as a product of unconstrained intellectual endeavour (Adorno, 1947, 1991). In Spracklen (2009, p. 34), I argue that Habermas resisted the overly pessimistic view of modernity in the writing of the Frankfurt school and returned to Marx to account for human progress, but that alignment with Marxism did not last:

> Habermas could have simply become another member of the Frankfurt School, aligning himself with critical theory's pessimism about the evils of modernity and the failure of rationality. Instead, he developed his own position on critical theory to account for individual freedom and agency within an optimistic view of progress taken from Marx (Finlayson, 2004). In doing this, he fell out with Horkheimer, and left the Institute before his post-doctoral thesis could be formally examined. But his brief connection with the Frankfurt School opened up other career opportunities in philosophy and sociology, and Habermas soon became an established Professor and member of the liberal-left West German intelligentsia. His alignment with Marxism was broken when he disagreed with the confrontational tactics of hard-line Marxist student activists (Finlayson, 2004), but he remained a key critic of capitalism and totalitarianism/fascism, and a defender of liberal democracy. By the 1970s he was also a public figure in West German civic life, supporting federalism and the enlargement of the European Union; but also, ultimately, criticising the unification of Germany and the creeping growth of nationalism and revisionist history (Muller, 2000; Traynor, 2008). For Habermas, the idea of Europe as a public sphere (Habermas, 1962:1989) was evidenced by its transcendence of national self-interest, and the establishment of a shared, civic discourse. This contrasted with the myth-making and self-serving stories of patriotism that, especially in Germany, resonated too closely with the far-right ideologies of the early twentieth century. Habermas' later political thinking, then, was a product of his earlier struggles with

authority and autonomy: Habermas attempts to defend reason and the philosophy of the Enlightenment and Truth (Habermas, 1998). Both, however, faced difficult challenges at the end of the twentieth century.

Habermas sees modernity as a construction of two competing drives, urges, spirits or rationalities, which I will discuss in more detail below. There is, firstly, the spirit of the Enlightenment (Habermas, 1984, 1989[1962]). This moment in Western history allowed individuals to escape the constraints of feudalism, traditions, prejudice and superstition. In the work of Thomas Paine, for example, there is a call for full equality and democracy in social life, what Habermas calls the public sphere. The public sphere was a construction of communicative rationality and action, the free will and the democratic agency of free individuals. It was exemplified by the free press and the coffee shop and the salon – spaces where educated, bourgeois Westerners could speak freely, listen to others and make up their minds about how to live, how to be fair, how to make laws, how to build better societies, how to understand the truth of nature. This public sphere created our lifeworld, the space where we make meaning and purpose, in which modernity was transformed. However, the potentiality of the lifeworld was soon endangered by two developments in high or late modernity, each a result of the rise of instrumental rationality. The first danger to the lifeworld came from the rise of the modern nation-state. Nation-states create systems that monitor behaviour and control citizens; these systems become processes of bureaucratic rationality, limiting our freedoms and promoting narrow, nationalistic ideologies and identities. The second danger comes from global capitalism, which reduces the meaning and purpose of individuals and practices in the lifeworld to a logic of profit and loss (Habermas, 1987).

Habermas remains an important champion of critical theory. His awareness of the problems in historiography, and his historical sensitivity, makes his account of the history of modernity compelling. He has a through grounding in philosophy and understands the difference between a reasoned argument and rhetoric: his entire academic career has been shaped by the Enlightenment tradition of German critical philosophers, from Kant to Heidegger. He does not simplify or distort history to make points in his political or sociological theory – a failing that undermines many sociological accounts of modernity. His awareness of the power of instrumentality aligns much of his theoretical framework with the other critical theorists of the Frankfurt School, and also aligns his framework with various post-Marxist accounts of modernity and structure, especially Gramsci's notion of hegemony. His account of communicative rationality enables us to see that there is a right and wrong way of being, a set of universal rules about how we live that allow us to fight for social justice. And his defence of critical scholarship allows us to be confident of our ontological and epistemological positions in an Academy where various critics have claimed there is nothing solid for us to hang our theories onto.

Habermas and leisure

Again, I need to refer back to my own work. In my previous two books for Palgrave Macmillan, I explored the meaning and purpose of leisure and attempted to make sense of leisure using the theories of Habermas. In the first book, *The Meaning and Purpose of Leisure*, I wanted to understand our contemporary leisure lives here in what we might call late modernity. I was interested in the paradox of leisure: how could leisure be two contradictory things, something that is about free choice and free will, yet also something that is forced upon us by the powers of modern society? I was not satisfied with the arguments made by other leisure theorists – that leisure could be reduced to free choice (the common-sense approach, what I called the liberal theory of leisure), or reduced to constraint (the structural theory of leisure). And I was not convinced by the arguments that leisure had become a chimera due to the shift to some place that was called the postmodern. Essentially, I was and still am a post-Marxist, looking to fight for social justice and wishing to preserve the idea of structure in the analysis of leisure, since structures are still evidently important in society. I was also, philosophically and politically, a liberal with faith in the uniqueness of humans to use reason to make their world a better place. Habermas offered me a way of reconciling the potential paradox and inconsistency in my own philosophical position and in my leisure theory.

In the first book, I use Habermas' key theory of two opposite rationalities to solve the paradox of leisure and save leisure theory from confusion and contradiction. I explore how Habermas' work on globalization, modernity and postmodernity can be applied to understanding the tensions within leisure studies about leisure's place in all these social turns. I explore how Habermas himself used leisure examples in his work. In his account of the development of the public sphere, leisure plays a key role in identifying the historical practices of various stages in Western history – the shift from the formal, elite leisure spaces of the

Greeks to the arbitrary and religious nature of leisure in feudal Europe, and the importance of informal leisure in constructing the public sphere in the Enlightenment (Habermas, 1989[1962]). Leisure examples are used by Habermas in his later work on the project of late modernity and the rise of instrumentality: the rise of popular culture as mass culture, the rise of television and cinema, and the transformation of leisure into an industry (Habermas, 1987, 1990, 2000). I then apply his theory to three of my own research projects to make sense of them.

In the second book, *Constructing Leisure*, I sketch out a philosophical history of leisure, which looks at different philosophies and uses of leisure in different historical periods, in different places. This book begins with a history of philosophy of leisure, exploring ideas on leisure taken from key philosophers. The historical account starts with the pre-history of humanity and the archaeological record. I suggest that although the record is slight, it is clear that our pre-historical ancestors found meaning and pleasure in leisure. I discuss leisure in pre-modern cultures and societies from around the world and identify nascent instrumentality at work in a number of places, especially the Roman world where leisure practices and spectacles were controlled by the State. Habermas is interested in the shift from pre-modern to modern, and my book follows his work on the shift in Europe from the feudal to the early modern. But I also look at the shifts to modernity in countries such as India, Japan and China, and the role of leisure in such processes. I argue – following Habermas – that leisure has an essential meaning and purpose for humanity, but that the communicative need for leisure is always in danger of being subsumed by the instrumental use of leisure by those who try to control the world. As I discuss in Spracklen (2009, pp. 89–90):

> Habermas was never a leisure scholar, and leisure was never the primary focus of his various theoretical and research programmes. Nonetheless, he was interested in the way in which leisure was used to establish key concepts in those programmes. Leisure played an important role in his history of the development of the public sphere. Leisure is seen as a form of behaviour that allows both communicative rationality to be nurtured, and as a form that works against the communicative, democratic discourse under the pressures of globalization, capitalism and State bureaucracies, He saw how leisure was privatized in the onset of Modernity. He argued that instrumentality was commodifying leisure and culture in such a way that made it difficult for individuals to find the political and

social space to think freely. For all these reasons, his work on leisure is important for leisure studies, and how research into leisure is maintained in the twenty-first century.

This Habermasian framework is one that has been undeservedly over-looked in wider theoretical debates about leisure (though see Scambler, 2005, and Morgan, 2006); yet it is one that offers a way of recon-ciling arguments from structure and arguments from agency without resorting to epistemological despair. Some of the absence of interest in Habermas is a product of the education leisure scholars generally receive as undergraduates: Habermas has never been widely referenced or used in classrooms in the United States and other English-speaking countries. This might be a product of deliberate choices over what to include in curricula, or the absence might be a subconscious lack of interest in German critical philosophy, which is connected with wider stereotypes about dry, methodical Germans (see discussion in Watson, 2010). I was lucky to be introduced to Habermas through studying history and phi-losophy of science. However, the work of Habermas has been strongly critiqued by a range of postmodernist and post-structural scholars, and this might be the source of leisure scholars' lack of engagement with Habermas: he has been accused of defending reason.

Post-structural criticisms of Habermas

In 'Modernity versus Postmodernity', Habermas (1981) begins by won-dering whether it is right to follow the postmodern turn that rejects the notion that we can really know anything to be true or real, and that rejects the notion that the Academy can equip people with the skills to make sense of the empirical world. Habermas proceeds to reject the fundamental claims of postmodernism. Firstly, he argues that there is no coherent theoretical framework behind postmodernism, only rhetorical flourishes and positioning. Secondly, he suggests that post-modernism is philosophically incoherent because postmodernists reject normative statements but fill their publications with clear, normative statements based on what they assume to be right and wrong. Thirdly, Habermas claims postmodernism is unjustifiably reductive and opaque in its ontology, failing to accept that things take place beyond the sym-bolic and metaphorical. Finally, Habermas argues that postmodernism cannot provide the tools to understand the everyday life we lead. He turns to post-structuralism in two lecture essays on Derrida published in *The Philosophical Discourse of Modernity* (Habermas, 1990, VII and XII, pp. 161–184 and pp. 336–367). In these two essays he questions whether Derrida's method of deconstructionism could ever be used to make sense of the social world. If everything is reduced to a sign in an infinite regress of signs and symbols, and no one has access to the right way of reading those signs, and no external epistemological framework exists to help us make normative judgements about how we read the texts, then all deconstructionism becomes is a game with no rules.

These essays generated distrust and cynicism of Habermas among postmodern and post-structuralist scholars, who countered that Habermas' belief in rationality and the Enlightenment project was a naïve ethnocentric belief in positivism, which had been strongly rejected by the postmoderns (Hahn, 2000). The idea that there could ever be a universal truth or a true picture of reality was a product of white, imperialist ideology: the Enlightenment was a product of white elites who supported slavery and made profits from exploiting workers – it was easy for white elites to claim that their particular point of view was true and that of non-white subalterns was false because white science was backed by white imperialism and hegemonic power (Meehan, 1995). The strength of the accusations made about Habermas led some of his defenders to fight back, and for a time in the 1980s there was a struggle in journals and at conferences between the Habermasians and the post-structuralists. This did not stop Habermas and Derrida from finding common ground in later years, and they worked together to try to show their more rabid followers that they both believed in an ethics of social justice and responsibility, and both were concerned about the rise of American hegemony and globalization (Borradori, 2004).

The accusations about the Enlightenment are a gross conflation of two separate issues – the failings of science and the whiteness of the Enlightenment philosophers – and stem from a general cynicism about science and reason that has taken root in popular culture and the Academy. Some of this scepticism about the Enlightenment comes from post-structuralist scholars such as Foucault, who shows how particular structures and ideologies were built into the medicalization of madness and the development of law and order in the Enlightenment period (Foucault, 2006). There is no doubt that science does not have all the answers, and there are limits to understanding. There is no doubt that a naïve faith in scientific progress has created many problems for modern society and our planet. But logic and rationality, theory and method, induction and deduction remain powerful tools that have helped us make sense of the world around us – and to improve society in a number of important and meaningful ways, from the advances in technology to the public health schemes that have reduced the chances of people dying. Philosophers of science recognize that scientific knowledge is partial and the ethics of science are not always fully considered, but that does not stop us from using the tools and methods and knowledge to seek better lives. Modern science may be a product of the West and the product of Westernization and secularization, but that has not stopped science and scientific knowledge from being used across the world. To be cynical of science is to be cynical of the knowledge and the assumptions and the models and the experiments and the training and the innovations that keep people flying in planes. To be cynical of truth in social sciences and the humanities is to reject the notion that we can build a better social world, one free from racism and prejudice. In this, then, I stand fully behind Habermas' defence of reason and its application to social sciences and humanities: lose sight of the ability to judge something to be right or wrong, true or false, good or bad, and all that is left is a relativistic discourse where being racist is as equally valid as being an ant-racist.

On the whiteness of Enlightenment philosophers, as I have discussed earlier in this book, it is a historically correct fact that the Enlightenment was a product of white, Western bourgeois elites. It is a historical accident that the ideas of democracy, equality and freedom and so on were given a chance to develop and spread around the world because of the circumstances of modernity on our planet. But the specificity of the structures and culture that spread these values does not make those values less universal: the whiteness of the founders of the Enlightenment does not make their ideas culturally bound by their whiteness. These ideas are universal to all humanity and are found in many cultures and societies across the world, pre-modern and modern (Habermas, 1990). By embracing the ideals of the Enlightenment, Habermas is not embracing an ideology that supports the hegemonic power of whiteness and the inequalities of history. Rather, he is embracing a belief in truth and social justice that challenges racism, inequality and inequities of power; a belief in communicative reason that will allow us to make sense of the ways in which modernity acts to construct dominant discourses of nationalism, Americanization, masculinity and whiteness. Rather than reifying white privilege, Habermas' defence of communicative reason provides a space in which such privilege can be challenged – when such privilege becomes identified with a form of instrumental rationality about 'race'. As I suggest in Spracklen (2009, pp. 67–68):

> In rejecting postmodernity, Habermas is explicitly setting out his defence of modernity as a project of reason, based on the

Enlightenment. As discussed in the previous chapter [of *The Meaning and Purpose of Leisure*], Habermas wants to be able to understand meaning, philosophically and in the social world, to enable us to make secure judgements about truth, reality, morality and justice. In defending modernity, then, Habermas is defending the lifeworld, where communicative action allows individuals the freedom to think rationally and make such judgements collectively. In the lifeworld of modernity, there is recognition of injustice and the constraints of instrumentality, and the role of the philosopher-politician seems to be a resurrected version of the role of Plato's philosopher-kings of *The Republic*: to guide the polity into making the right decisions about itself, so that people live good lives according to reason and justice. This is one reason why Habermas is so angry with the postmodernists: in setting up false equivalents and chimeras of relativism, they seem to allow injustices and constraints to continue in the name of diversity (Habermas, 1998). So in the Balkan conflicts of the 1990s, misguided notions of impartiality stopped Western governments condemning the siege of Sarajevo because they did not want to be seen as pro-Muslim or anti-Serbian. And the basic, communicative human rights of women have often been compromised by Western politicians scared of condemning misogynist cultural practices in other countries.

Habermas argues that both postmodernism and postmodernity need to be rejected because they are chimeras of criticality. The former notion claims to introduce something new to our understanding of the world by questioning truth, but there is nothing new about scepticism: critical reasoning demands scepticism about knowledge claims and the ontological status of the things we study. But being sceptical about the limits of knowledge and certainty does not mean we abandon reason for relativism, indeed it is an erroneous extension of the argument if we think that evidence of uncertainty is evidence that nothing can ever be known. Postmodernity might be more acceptable to Habermas as it denotes a period of human history, but this he rejects altogether. There is too much of the postmodernism in postmodernity for it to be a useful descriptor of contemporary society: things have changed in the world, especially in the West, and modernity might be associated with the previous century, but there is no hard evidence that the structures and societies that are identified with modernity have disappeared anywhere in the Western world. Postmodernity, like postmodernism, is an intellectual exercise in denying truth to the other side of the debate by shouting very loudly and covering one's ears when the other side is speaking – postmodernism is a pseudoscience like Creationism, pushed onto the curriculum by rhetoric and finger-pointing, but logically incoherent and lacking any real evidence for its existence.

Communicative reason

The strength of Habermas' account of the emergence of the lifeworld is in the role played by communicative rationality in its construction and maintenance. Communicative reason is the glue that binds the lifeworld of individuals together, the thing that turns a collection of individuals into a social world that is greater than the sum of its parts. The lifeworld itself is the most famous metaphor Habermas created. It is in one sense a mere description of the social world created by communicative rationality, the culture or society in which individual actors use their agency to debate the shape of the social world: it is the place where the public sphere forms and is acted upon. But the lifeworld is a lifeboat, a ship holding our humanity afloat in a sea of uncaring instrumentality. The lifeworld is an island that is colonized by instrumentality. The lifeworld is the metaphor in which culture emerges through the discourses and detours of everyday life. As Blumenberg argues (2010, p. 96), thinking of culture and the metaphor of journeys:

> There are an infinite number of detours from point of departure to destination, but only one shortest way. Culture consists in detours – finding and cultivating them, describing and recommending them, revaluing and bestowing them. Culture therefore seems inadequately rational, because strictly speaking only the shortest route receives reason's seal of approval. Everything right and left along the way is superfluous and can justify its existence only with difficulty. It is, however, the detours that give culture the function of humanizing life. In the strictures of its exclusions, the supposed 'art of life' that takes the shortest routes is barbarism.

Blumenberg's metaphor of the detour might be critical of rationality, but this rationality he refers to is instrumental, the bottom-line of the modern world. His detours that make culture are the discussions and discourses that are the basis of the Habermasian communicative rationality. When we ignore the shortest path we become human and use our communicative powers to not only shape culture but also shape leisure. Our ability to be free to think and contemplate as human beings is constrained by the demand from hegemonic powers that we consider the shortest routes, but when our rationality is not constrained we have the ability to be fully human. In such situations, we express our humanity through the creation of culture and the use of leisure: things that hold no monetary value, but things that bind us to our fellow humans in our lifeworld. As I suggest in Spracklen (2009, pp. 48–49):

Communicative rationality is emergent, contingent on actors and action, and dependent on consensus over the hermeneutics of language. Furthermore, communicative action takes place in what Habermas calls the lifeworld, which 'comprises a stock of shared assumptions and background knowledge, of shared reasons on the basis of which agents may reach consensus' (Finlayson, 2004, p. 52). The lifeworld shares many similarities to Cohen's symbolic community (Cohen, 1985), or perhaps Bourdieu's habitus – field nexus (Bourdieu, 1979:1986, 1991). All three are based around the notion of shared principles, discourse, agreements on meaning, and the objectification of the social. But there is one difference between the lifeworld and these other imaginary constructions of belonging: the lifeworld is the totality of our social world, not just small set of our ontological environment. Habermas's concept owes its origins to the work of Karl Popper, who tried to make a distinction between the worlds of physical objects, mental states and cultural products (Thompson and Heald, 1982). The lifeworld is the world of the latter, a symbolic construction of the first two worlds. Ideally, the lifeworld is constructed out of communicative rationality, which leads to principles of progress, fairness and ethics being part of that world. So Habermas argues in his analysis of discourse ethics and morality: the lifeworld is where reason and discourse shape laws about equality, for example, which can be seen in the way in which racial discrimination has been slowly challenged and criminalised in many liberal democracies (Habermas, 1983:1990, 1991:1993). There is, then, a simple seductiveness about the theory of communicative action and communicative rationality. The latter is an ideal of objective, unfettered reasoning, drawing on structural realist ideas of meaning and language, which in turn inherit older notions of rationality from Kant and Plato. The former is the way in which the latter is acted out in discourse: free from constraint, democratic, fair and consensual in its rules and outcomes. In an ideal world, the lifeworld would be built solely from such stuff, and progress and truth would be assured to all those in it.

If the lifeworld floated freely, it would be possible for racism and prejudice to be overturned. In a public sphere where all humans dis-coursed equitably, and enjoyed equal access to knowledge and power, there would be a communicative space that celebrated our common humanity, our differing narratives of identity and belonging, and our individual differences. Leisure and culture would be things that are per-manently in flux, hybrids made and un-made to suit our collective and communicative needs. We would be able to construct better ways of managing society, there would be a wholesale, radical change in how we distribute opportunities and access, as the focus would shift to ensur-ing equality of outcomes. We would create games and systems in ways that reflect the rules of social justice identified by Rawls (1971): ensur-ing that everybody starts out with the best possible chances in life, and that individuals and groups are not allowed to concentrate power that can be used to rig the system in their favour. In this sense, the Habermasian lifeworld begins to sound like a utopian dream, but these consequences are based on the communicative rationality from which the lifeworld is constructed. Like in the blind guessing at the heart of Rawls's social justice model, people have to construct a lifeworld that will ensure they get the best out of life without the certain knowledge they belong to a hegemonic elite that will maintain that hegemony in years to come. Communicative rationality makes people 'hedge their bets' and trust more on finding equitable, democratic solutions, rather than hoping their elite status will continue to survive without the power of instrumental rationality. Of course, in real history, elites have relied on instrumental rationality. Such instrumentality has distorted the real history of the lifeworld and continues to work to keep people with power in possession of that power, as I will explain next.

Instrumental rationality

Instrumentality is the rationalization of reason, and the reduction of communicative rationality to ways of thinking, which are constrained by models of economy and bureaucracy. Thus instrumental rationality is the rationality identified by Weber, as Habermas recognizes (Habermas, 1990, pp. 1–2):

> [Weber] described as 'rational' the process of disenchantment which led in Europe to a disintegration of religious world-views that issued in a secular culture. With the modern empirical sciences, autonomous arts, and theories of morality and law grounded on

principles, cultural spheres of value took shape which made possible learning processes in accord with the respective inner logics of theoretical, aesthetic, and moral-practical problems. What Weber depicted was not only the secularization of Western culture, but also and especially the development of new modern societies from the viewpoint of rationalization. The new structures of society were marked by the differentiation of the two functionally intermeshing systems that had taken shape around the organizational cores of the capitalist enterprise and the bureaucratic state apparatus. Weber understood this process as the institutionalization of purpose-rational economic and administrative action.

Instrumentality is the construction of quasi-scientific models of economy and society, something which Weber and Habermas both attack for labouring under a false assumption about how science works with models. To understand the philosophical debate about the nature and role of models in science, it is necessary to understand how the debate about how scientific theories represent has developed since the time of the logical empiricists (Giere, 1999). According to the proponents of the Received View, a theory is constituted by a set of axioms (the calculus) and a corresponding set of rules that relate the phenomena to the theoretical laws and yield a *partial interpretation* of those laws (Carnap, 1939). In the Received View, then, a model can be seen as just another interpretation of the theory's calculus. As Braithwaite argues, the use of a model '(apart from its psychological function in illustrating the theory) is to approach indirectly a difficulty which the contextualist attacks... clear-sightedly by talking of his calculus' (Braithwaite, 1962, p. 231), that is, a model is an unnecessary complication that may have some heuristic value (Carnap, 1939) but that may at worse confuse and mislead. This view of models was dismissed by the likes of Hesse (1963), who argued that models in science can provide more enlightenment of the phenomena than the theory alone. Hesse developed the concept of the neutral analogy used in model building. An analogy, argued Hesse, can be viewed as positive (we can see there is an analogy), negative (we can see there isn't an analogy) or neutral (we don't know whether there is an analogy or not). By making a neutral analogy of properties in model building, we can then use the model in a heuristic fashion (Hesse, 1963). For example, the billiard ball model of molecules in a gas will have a number of neutral analogies about the behaviour of billiard balls, which may or may not be applied to the behaviour of models in a gas, but which in themselves provide a basis for further investigation and explanatory accounts of the behaviour of the molecules in a gas. Another criticism of the Received View of models was developed by Achinstein, who showed that scientific practice in developing and using models was far more complex than Hesse's account of models as mere interpretations of the calculus (Achinstein, 1968). This view was also shared by Suppes (1957) in his explication of the Semantic View (or, as Suppes calls it, the set-theoretic view). Models are used in a number of ways by scientists. Suppes argues that the diversity and richness of models in science is not captured by the Received View. This is a weakness of the entire Received View account of the relationship between theories, models and phenomena in science. Suppe developed the work of Suppes by pointing out that theories are constructed from a cluster or hierarchy of models, with scientists using models at a number of levels to connect theory and phenomena. According to Suppe (1977, p. 221, emphasis added):

> Theories are not collections of propositions or statements, but rather are *extralinguistic entities* which may be described or characterised by a number of different linguistic formulations. This observation does not show that an adequate understanding of theories cannot come from an examination of the linguistic formulations of theories, but it does suggest that such an approach is likely to result in a distorted picture of the nature of scientific theories.

So for Suppe, the Semantic View overcomes the weaknesses of the Received View by allowing theories to be extralinguistic entities, which do not, therefore, become tied to a particular language. Extending the Semantic View, Van Fraassen has argued that clusters of models are embedded in a common state space through a formal isomorphism (Van Fraassen, 1980). Giere (1988) has argued differently, eschewing formal accounts for a concept of similarity between models and phenomena. Applications of the Semantic View to account for theorizing in science have been made through a number of detailed case studies (see Lloyd, 1984), though not without critiques from some philosophers of science who have argued that the formal structure of the Semantic View cannot capture the complexity of the relationship between phenomena and theory in, for example, certain qualitative life sciences (Horan, 1988). Cartwright has also attacked the 'top-down' view of model building she claims is central to the Semantic View, and has argued that modelling in scientific practice occurs just as often from the phenomena, rather than from theory. For Cartwright, models are the 'blueprints for laws' (Cartwright, 1997), which in turn are idealized laws that can only ever be true under ceteris paribus conditions (Cartwright, 1983). Morrison has gone on to claim that models have autonomy from theories, in effect models can take on a functional independence – that is, models function partially independent of the theory to which they relate – in the way they are used to explain and account for phenomena (Morrison, 1999). Both Cartwright and Morrison believe that these aspects of models undermine the Semantic View, which, as Morrison claims, merely 'suggests that we need models where we want to fit the theory to concrete situations' (ibid., p. 42).

Instrumentality comes from the application of modern models of rationalization, industrialization and securitization to making sense of society – assuming that society works like a positivistic notion of scientific reality, with hard-wired rules of epistemology based on the logic of mathematics. In this worldview, social science accounts of agency and structure are replaced by models of efficiency. Instrumental rationality is a simplification of reasoning, a reduction of complicated discussions about any problem to simple *a priori* assumptions about the limits of the logic. Instead of allowing issues to be critically debated and assessed from all angles, instrumentality builds models that have their own internal rules that serve the interests of those in power: so the logic might say that all possible actions have to be judged for their economic benefit, or all possible actions have to be judged for the power they give the rulers over the ruled in a nation-state. There are two strands to instrumentality in modernity: the strand associated with the rise of economic liberalism and global capitalism; and the strand associated with the rise of modern nation-states. Like Weber, Habermas argues that such instrumentality only appears at a particular moment in high modernity. However, instrumental rationality appears in any society where power relationships allow the ruling classes to exploit and adapt the models of thinking in a systematic way – so there are echoes of instrumentality in places like the Roman Empire, even though Weberian rationalization only becomes widespread and hegemonic in industrial societies (Spracklen, 2011). For Habermas, instrumental rationality has increased over the last 150 years, and, in the West and in many other parts of the world, it is now so dominant that it makes it almost impossible to think and act in ways that are not instrumental. The power of instrumentality is almost total; it is now hegemonic in nature, with only a small part of the lifeworld left above instrumentality's floodwaters. This is why I argue in Spracklen (2009, 2011) for the importance of leisure in our lives: it is one of those activities that has the potential to be more communicative than instrumental. There are many ways in which leisure has become subject to instrumental rationality – the commodification of leisure, the rise of the leisure industries, professional sports and the sports-media complex, and the rise of popular culture as entertainment – but leisure in its purest sense is a communicative action freely chosen, an activity that has the potential to provide a communicative space in which the lifeworld can be saved from instrumental rationality.

Economic instrumentality is created by the demand of capitalists to reduce all transactions to ones that can be measured in cash value. The reduction of things to commodities and the pricing of everything then becomes a political imperative as modern elites realize their power is dependent on profit-maximization – the model of monetization becomes the only allowable way of settling decisions about what activities take place in late modern society. We can see this occurring in leisure in the West, and in most other parts of our social lives. In education, for example, the meaning and purpose of learning is reduced to a cost–benefit ratio that tells students, and their parents, studying a particular course if the student would get a well-paid job. So education has become a factory selling products that 'guarantee' wealth creation and economic success, and courses that do not have an obvious employability route for students become marginalized or get cut: philosophy is replaced by business studies, leisure by physical education. The whole purpose of education – its worth in making young people humans, its value in expanding horizons and getting young people to think communicatively and critically, its use in Bloomenberg's sense as a detour – is lost in the instrumentality of taking their money and promising them high-status jobs.

Systematic instrumentality is an outcome of the creation of modern nation-states. These states need to be able to count their citizens so that they can monitor their behaviour, take taxes and apply laws. Nation-states impose on their citizens invented traditions of national identity, loyalties that are used to ensure obedience. This nationalism is partly about generating blind support for the hegemonic elites who control nation-states. But it also serves the function of allowing nation-states to legitimize all manner of actions, interventions and systems that function to keep citizens in positions of inferiority. Systems instrumentality borrows from economic rationality the concepts of efficiency and performance management, and turns governance into bureaucracies of control and command, where large numbers of managers and processes emerge to monitor other managers and others processes. This is justified by the logic of national pride, subjecting all workers in the State bureaucracies to the pressure of working harder for the same salaries in the justification that national success is equated with increases in economic efficiency. The citizens at the bottom of this systemic instrumentalization of the State are encouraged to take sides in political debates that are constrained by the extensions of the State machinery in the media, so that they feel free, but in fact the machinery of the State works to limit their freedoms and funnel power up the systems to the elites. This model is then repeated in any large organization in modern, Western society. Again, education demonstrates the rule. Schools have increasingly become controlled by the central education ministries in countries such as the United Kingdom, dictated by politicians seeking to manipulate education for short-term political gains. And in higher education, the freedom of academics to debate issues and to become involved in the management of their own institutions is limited by increasing surveillance of activities, a proliferation of forms to fill in and a reduction of power on decision-making committees for academics and union representatives.

Instrumental whiteness

Habermas does not write directly about whiteness, but he does discuss the instrumental logic of nationalism, ethnocentrism and the construction of myths of racial purity. For Habermas (2000), exclusionary and ethnic-based nationalism is the ideology that leads to the emergence in the nineteenth-century of modern nation-states out of the multi-ethnic empires and states of the West. The elites of these modern nation-states impose invented traditions and racialized logic that denied plurality and multiculturalism in favour of racialized myths that protect the power of the new elites: nation-states are bound by shared languages, shared origin stories and shared blood. This racialized nationalism continues to shape identities and politics well into the twentieth century – and even after the millions of lives lost in the two world wars, post-colonial geopolitics was shaped by the instrumentality of homelands, flags and anthems, and other artefacts associated with imagined communities (Habermas, 1989[1962]). Racialization is at the heart of late modern nationalistic ideologies in this present century, and for Habermas this instrumentality is one of the most pernicious threats to the development of a truly transnational democratic system (Habermas, 2000). Such racialization might, then, seem at odds with the instrumental logic of global capitalism. If economic instrumentality is about open-ing the markets across the world for exploitation and profit-making, then racialization should be counter-productive to that logic: the market should demand free movement of people as well as capital, and the mar-ket should not be influenced by the colour of the skin of the buyer or seller. But instrumentality is about the preservation of existing power, the hegemonic relationships established by the accidents of history that gave rise to the contemporary world. Economic instrumentality is dominant because it serves the purpose of the existing hegemonic powers to open up new markets, access cheap resources and make them-selves richer at the expense of the people exploited by the logic of the markets.

We can follow Habermas' theoretical framework to its own logical conclusion. Instrumentality serves to constrain communicative free-doms and to maintain the power of hegemonic elites who have his-torically been successful in acquiring the keys to the engines of modern nation-states. Instrumentality is the key to the construction of exclu-sionary identities and groups, the racialization of people without power and the wishing away of the racialization through the magic trick of hegemony. Instrumentality serves to construct and maintain the Western elites who hold hegemonic power: it serves to construct and maintain their whiteness through the constant dynamic of Othering and whitewashing of identity formation. Instrumental whiteness then is not only an accident of history, a description of the white people who still form the Western hegemonic elites; it is a form of logic that operates to hide the work of racialization and to legitimate the continuing power of the West in the globalizing world society in which we all live. Instru-mental whiteness is hegemonic whiteness, the whiteness that adapts its own symbolic boundaries and myths of purity to maintain its power, the whiteness of the new global aristocrats: the senior managers and bankers, the Western diplomats and politicians, the executives involved in the transnational media corporations, who have all been adopted into the ranks of the older Establishments (the WASP societies of America, the gentleman's clubs of London in England). Instrumental whiteness serves to protect the old imperial interests of Europe and the United States, which are now wrapped up in the post-colonial military and cul-tural hegemonies of the West (Habermas, 2000). It is an elite whiteness that will use crude nationalism and racism when it serves the purposes of the elites (for example, through stirring up poorer white communities to commit hate crimes against minority ethnic groups), and which will appear to be more tolerant of pluralism and hybridity when it needs to be more invisible in its use of its hegemonic power. Like the notion of the free market or the ethnically pure nation-state, whiteness has no real ontological status beyond the myth-making of the instrumental logic (it could be instrumental blueness, if the accidents of history had led to the dominance of blue-eyed people – who identified as such – at the point of high modernity, when instrumentality becomes the dominant mode of rationality). But the impact of instrumental whiteness is the continued racialization of everyday life, including leisure. Of course, although this whiteness is instrumental and hegemonic, there is still potential for resistance through the use of communicative rationality and action. And although instrumentality is a product of the Enlightenment, we must not reject the Enlightenment outright. The existence and survival in the Academy (and beyond it) of counter-narratives of 'race', predicated on communicative rationality, shows that the Enlightenment – while flawed in history – remains a durable ideal of free inquiry.

Conclusions

Communicative rationality, the way of thinking that comes from our humanity and our identity as social animals, leads to the establishment of the public sphere and the lifeworld, in which our leisure choices help us construct civic (and civil) society. In modernity, each of us belongs to the lifeworld – that part of modern life that emerges from the public sphere. All of us have the education and the reasoning to be able to think for ourselves, to construct with others a mutually beneficial world for our culture and society to thrive. In the lifeworld, we are free to choose leisure activities that give meaning to us – these could be solitary activities bounded by social norms, or they may be social activities that bring us together with others to share meanings and discourses. What counts is the level of freedom such activities provide us, and the level of flexibility to change and adapt those activities to suit our decisions and deliberations. Being able to think for ourselves, being able to find solace in leisure, to be able to choose to make our own meaning from leisure, these are fundamental markers of humanity. But that lifeworld is in danger from the approaches of instrumental structures and organizations – the State, bureaucratization, commodification, hegemonic capitalist power and globalization – that threaten to colonize the lifeworld and destroy it altogether. Sports instead of being about communal effort and individual satisfaction become professions, with elite athletes transformed into brands to make millions of dollars for sportswear companies. Tourism becomes just another industry taking people's money and selling them false dreams, when it could have been based on personal growth and mutual respect. Culture becomes a factory assembly line making new shiny singers and film-stars who will make us buy their owners' products in the false belief that we are consuming art. And leisure becomes a site where instrumental whiteness is perpetually (re)produced in a dynamic of racialization and hegemonic subterfuge.

5
Whiteness and Popular Culture

Introduction

The *Harry Potter* franchise is a phenomenon that became, in the early part of the twenty-first century, a globally known brand of later modern popular culture (Gupta, 2009; Kidd, 2007; Lee, 2012). The franchise consists of best-selling novels packaged in different covers for children and adults, movies with multi-million-dollar budgets and multi-billion-dollar receipts, toys and games and collectables, fan communities in the net and in the real world, theme parks and branded events, tie-ins with fast-food companies and other food and drink producers, and packaged and themed tourism opportunities (see Lee, 2012). The franchise started with a children's book, *Harry Potter and the Philosopher's Stone*, written by Scottish author JK Rowling (1997). In this book, an English boy called Harry Potter, who is living a hard life as an orphan with his relatives in a very English, very white suburb (from the cover of the book it is clear Harry is white – as indeed he is when portrayed in the films), finds out his dead parents were wizards and he has been enrolled in a special school for magic, Hogwarts. The school is a pastiche of the classic British public school, with its boarding rituals, with its obsession with houses, formal dinners, strong discipline and school sport. Harry makes friends with Ron Weasley and Hermione Granger: Ron is definitely white as he has ginger hair and freckles, he is also from a wizarding family; Hermione is not described as being white but is white in the films, and she comes from a Muggle family, parents with no magic 'blood'. Through the series of books, Harry and his friends fight on the side of good against evil. The schoolchild heroes are supported by Professor Dumbledore, the old headmaster (again, a white character on screen), and a number of other teachers and wizards. Against them are ranged the forces of the evil Lord Voldemort, and Voldemort himself turns out to have been a former pupil at Hogwarts called Tom Riddle.

  The first two books are relatively simple and draw on the style and settings of a range of British children's authors, from Enid Blyton to CS Lewis. As the series progressed, and when the film franchise started to eclipse the books in terms of global impact, the storylines became more considered, more adult in theme. Rowling developed or introduced a number of minor characters who were black or recognizably from some other British minority ethnic group. The evil characters in the franchise started to talk about keeping wizarding blood pure and of keeping Muggles and mudbloods (what they called people who had non-wizarding heritage) in check. On one level, then, Rowling and the writers and directors of the movies were trying to use the popularity of the franchise to make a salient point about the way racism and fascism works in the real world. But on another level, the franchise problematized racialized thinking and normalized the particularly elite version of white, Western popular culture that underpinned it. The problematization of racialized thinking occurred throughout the book and through the films, whenever the concept of Muggles and family and magical powers was raised. Harry owes his magical powers to his parents – he has the advantage of his blood. Ron is initially a hopeless wizard – but his is a well-established wizarding family and again he becomes good. Hermione's parents are not wizards but she has still been sent away to wizard boarding school – the underlying assumption being that wizard school is the best place for her, away from the naïve folk who somehow ended up with a daughter who could do magic. The normalization of hegemonic whiteness – and a particularly elite version of it – is more evident in the films than the books, but is present across the franchise. In the films, every single major character, good and evil, is white: Harry, Ron, Hermione, Dumbledore, Snape, Hagrid. Nearly all the minor characters are white, with just a handful of token non-white faces in supporting or unspeaking roles. All the teachers and most of the characters speak in posh English accents and have the calm demeanour of junior officers in a Rudyard Kipling poem: the working-class Weasley family and the Scottish caretaker Hagrid are stereotypically stupid and prone to emotional outbursts. Part of the franchise's global appeal is precisely this quaint and archaic Britishness – the stories construct a Great Britain that people in other countries still like to think might exist, one fixated with class and hierarchy, tradition and ceremony, where tea is served hot with milk and sugar and cake and biscuits, where steam trains take you from fog-filled London streets to heather-clad mountains in Scotland in a matter of seconds. But the success of the franchise is also due to the dominance of Western, white popular culture in late modernity, and the way in which such representations of whiteness are still sold as normal in cinema screens and television sets across the globe. Consumers of the franchise in India might get excited when the Patel sisters are given a second of screen time, but for most Potter fans the attractive strangeness of the Potter world comes precisely because it does not represent their complexity and diversity.

This chapter will focus on whiteness and popular culture. The chapter will continue with a brief review of the existing research literature on whiteness and popular culture. The rest of the chapter will be divided into three sections, which will explore whiteness in particular contexts of popular culture and its relation to leisure. The first section of the main part of the chapter will focus on whiteness in television, drawing on and using examples from popular American TV shows syndicated worldwide, such as *Friends* and *Star Trek*. It will examine how leisure is used to construct whiteness within the shows and how watching the shows teaches consumers to be white. The second section of the chapter will look at whiteness and fantasy films and online gaming platforms such as *World of Warcraft*. It will show how these films, gaming platforms and video gaming can be used to counter instrumental whiteness by allowing some communicative space to subvert and resist dominant discourses – but such communicative spaces are rarely used by fans and gamers. The third example of whiteness and popular culture will be an exploration of popular literature: glossy magazines aimed at mass markets (male and female) and bestselling books (genre novels and non-fiction lists). It will be shown that instrumental whiteness, like hegemonic masculinity, is fully dominant in this part of popular culture, and leisure choices inevitably reproduce such whiteness at the expense of other identities.

As discussed in Chapter 3, theorization of whiteness in studies of popular culture is well established in North American sociology and cultural studies, though most studies of 'race' and popular culture focus on racism, blackness and Othering – with whiteness being the metaphoric shadow around which the theorizing and debate take place (Frankenberg, 1997). The discussion of whiteness in American studies of 'race' and popular culture is predicated on the particular history of racial politics in the United States, the shadow of slavery and ongoing racism and the ongoing struggles over popular culture in that nation between liberals and conservatives. Theories of whiteness and popular culture exist in the work of scholars working across the Atlantic, ones working in a post-Marxist paradigm, such as Stuart Hall and Paul Gilroy. Hall sees popular culture as a place of potential resistance to capitalist power, through the agency afforded to identity-making in subcultural scenes, but popular culture is also something that is constantly co-opted by capitalists and governments in their efforts to keep control of the masses (Hall, 1993). In *Between Camps*, Gilroy (2000, pp. 30–31) discusses the role of commodification, Americanization and the creation of a globalized popular culture (of American films, Americanized media and brands, and Americanized markets typified by the spread of Disney, Fox, Nike and so on) in the translation of particular ideologies of whiteness:

> Problems of compatibility and translation have been multiplied by the globalization of culture ... Certain common features, like the odd prestige attached to the metaphysical value of whiteness, do recur and continue to travel well, but they too will be vulnerable to the long-term effects of this crisis.

For Gilroy, then, whiteness is something that serves as an absolute metaphor for Western political and economic power over the construction and direction of modern culture – which is the direction of commodification. Social and cultural theories of commodification draw on Marxist accounts of capitalism and the loss of individual power and agency. Commodification is an extension of the effects of modern capitalism on society, and the establishment of unequal power relationships in rigid social structures. However, one does not necessarily have to be Marxist to recognize the fact of commodification. Liberal sociologists influenced by Weber's concept of rationality in modernity will see that such marketization and bureaucratization of social relationships are a consequence of the loss of the traditional and the rise of the rational. Functionalists influenced by Talcott Parsons would also recognize the increasing commodification of the world, especially the commodification of the social and cultural values we might place on some function or other in an ordered, modern society (such as the replacement of Christian charity in the West by a culture dominated by a model of individual gain). For Weberians and Parsonians, the process of commodification driven by modernity is just as real, though the ability to resist its effects is still given prominence.

On the structuralist side of sociology and social theory, the Frankfurt School has had the biggest influence on theories of commodification. Adorno (1947, 1991) was deeply pessimistic about the impact of modernity and modern capitalism on culture. He believed that the creative spirit of art, the genius of the beautiful and the human- ity of high, elite culture were all in danger of being lost by the commodification of modern society. For Adorno, the invention of the radio and the creation of the popular music industry removed human expression and appreciation from music, replacing them with passive consumption of ersatz (fake, but made to look real) pop and jazz. What was happening to music and high culture was happening in every part of human society, argued Adorno. Humans were becoming pas- sive consumers of material goods, living in a world where everything could be bought and sold, including humanity itself. This fear of the tide of materialism, economic rationality and modern capitalism is also present in the work of Jurgen Habermas. Habermas's theory of the state of modernity relies on the fight to stop a tidal wave of what he calls instrumentality – commodification and materialism – sweeping human agency away. Unlike Adorno, Habermas does not believe that complete commodification is inevitable, but he does believe it is happening and it is difficult to stop.

Gilroy builds his account of whiteness, 'race', commodification and power on Foucault. Foucault argued that the capitalist phase that emerged out of the Enlightenment, and in which we still live, objec- tifies and commodifies the human body. His concept of embodiment came from his reading of what he called the genealogy of madness and hygiene in his key early works *The History of Madness* (re-published 2006) and *The Birth of the Clinic* (1973). In the former, he demonstrated that the status of madness as a physical problem or disorder was a function of the increasing power of regulation, capitalist economy and science. In the latter, he showed how the role of the expert medical profes- sionals (and their appearance at the end of the nineteenth century) turned traditional relationships between the doctor and the patient into routinized, rationalized, commodified systems of control and coercion. What matters for the modern State, for Foucault, is the control of its citizens, the removal of those who are not productive workers and the policing of civilized behaviour. Through the process of embodiment in popular culture of things such as 'race' (blackness and whiteness) and gender, individuals learn how to read the wishes of the State, of mod- ern capitalism, into the control of their own bodies through policing illness, body size, appearance, depression and well-being; this embod- iment makes us into good citizens and makes us relatively powerless against the manipulations of our rulers.

What all these social and cultural theories of commodification have in common is an acceptance of the historical fact of commercialization.

This is the simple trend of turning our everyday lives and lived culture – our preferences, tastes and networks – into commercial, profit-driven, profit-making enterprises. Such a turn to commercialization is dependent on its sustained impact on the tools, machinery and management practices of modernity, but it is present in all periods of history where market economies have existed. In the Roman Republic, for example, private wealth dictated whether a free man was a senator, a member of the equestrian order or one of the lowly 'plebeians'. Private wealth could be inherited as well as earned in economic exchange, so men who wanted to move up the social ranks chased patronage as well as becoming speculators. Commercialization – the turning of everything into a market – seems to be a particularly human process: it is the nature, extent and pace of commercialization that is troubling in the modern world. The logic of the market has spread from the public sphere to our leisure lives.

Whiteness and television

Television has an enormous influence on how individuals see themselves and the world around them. Despite the rise of the Internet, television watching remains a dominant leisure activity in most of the world, and is one of the most important leisure activities in late modernity. When people in this late modern period are not watching television, they are reading about it in newspapers and magazines, chatting about it online or talking about it with their friends. For most Westerners, television has become the companion to their mornings and their evenings, and television viewing rates are counted in hours per day. The large proportion of our lives spent sitting down and watching the television is condemned by a range of pundits and academics across the political spectrum: from conservatives who accuse television of encouraging immorality, stupidity and sedentary lifestyles to radicals who accuse television of keeping people in thrall to capitalism (Bourdieu, 1998). Television in turn has been heralded as something particularly modern, which at its best educates and frees individuals from oppression (Fiske and Hartley, 2003). Television consumption for Westerners has become more privatized and individualized in this century, compared to the communal television-watching norms of the last century: individuals might watch television programmes in their own bedrooms instead of sitting down with the family, or they watch content online or on their computers (Andrejevic, 2008). Westerners increasingly pay for watching television programmes on specialist digital channels or as legal downloads – and they increasingly interact with other television fans online on various social media sites. In other parts of the world, television is more of a social activity, watched in front rooms or in public leisure spaces.

In almost every country, American television programmes dominate the airwaves. The dominance of American production companies and transnational media corporations with important American connections (for example, Fox, Disney or Universal) ensures American productions get sold in packages and distributed widely. The prevalence of American imports on the television networks of other countries has been notable since the 1960s, but has grown even more with the decline of television and film industries in other parts of the globe. This Americanization of television works to normalize whiteness and to trick viewers into thinking particularly American or Western forms of hegemonic whiteness are somehow part of a shared, universal human experience.

The well-known science-fiction franchise *Star Trek* serves as a good example of all of these trends (see Hark, 2008). When it was first aired in America the original series conformed to the expectations of 1960s American television drama: the captain was an all-American white boy, men were space captains and the United Federation of Planets protected and promoted a particularly American form of liberty, individual freedom and capitalism. The makers of the show did attempt to subvert the expectations of their owners and sponsors – so the show included women as crew members (though in very short skirts), and black people, and there were shows that used racism and discrimination as having no place in the future, but the show was careful not to push this liberal agenda too far. The show also included an alien as a leading character. Spock was described as half-human, half-Vulcan, and the writers essentialized his hybrid nature by stressing the differences of biology and character between the two 'races'. Despite his pointed ears, Spock still passed as white. The show was syndicated in America and sold around the world, gaining a massive fan-base that demanded more space adventures.

At the end of the 1970s the original crew returned in the first of a series of movies, and in the 1980s a new series – *Star Trek: The Next Generation* – was launched, which featured a new ship and a new crew, exploring the galaxy some years after the first crew. Of the ten main characters over the span of the series, eight, including the captain, were white. There were two black actors in leading roles, but one of them, Michael Dorn, played an alien Klingon called Worf. The show tried to distance itself from the expectations of American viewers, and was very successful at challenging prejudice – but the show developed a mythology about alien species that gave them stereotypical biological, psychological and cultural characteristics. Furthermore, the show's weekly adventures still frequently involved the second-in-command, rough and heroic white man Will Riker, acting like a cowboy in an old Western movie; or Captain Jean-Luc Picard drinking Earl Grey tea, sitting down calmly and solving the problem with all the restrained power of a British chap playing cricket. It was only at the third iteration of *Star Trek* (the 1990s series called *Deep Space Nine*) that a black human became the main character of the show, when Avery Brooks played the station commander Benjamin Sisko – and even then, he was not promoted to Captain until the third season. Deep Space Nine broke with the genre convention of the time and started to allow its characters to grow season by season, alongside long story arcs plotted over the life of the show. This switch from 'monster of the week' to 'soap opera' allowed Sisko's family to be carefully explored, and his relationships to develop. The writers of the show included sympathetic portrayals of Sisko as a struggling single father, as a black man struggling to make sense of his relationship with his own father, and through various plot devices the writers used Sisko to explore the racial politics of twentieth-century America. By the time the next series in the franchise came along (*Star Trek: Voyager*) at the end of the twentieth century, the captain was back to being a white person, albeit a woman, and the writing was more formulaic again. What these shows collectively demonstrate is the power of American media in late modernity, and the cultural hegemony of American ideologies of meritocracy and individual liberty – all the versions of this franchise depend on alien 'races' learning the lesson that individuals must have the right to choose their lives and be the best they can be; and the white men in the Federation uniforms show the aliens how they can try to do this.

The comedy show *Friends* has been another American product that has gained a place in popular cultural imaginations, being sold worldwide and generating an enormous fan-base (see discussion in Barker, 1999). The show sells to the world an image of white, bourgeois respectability, a cultural positionality that favours friendship and individual freedoms over other ways of being. The show's main characters are all white, young adults starting careers and searching for love and happiness. They live in fancy apartments in the middle of New York, spend their leisure lives drinking coffee at their local coffee shop, and have fancy materialist habits: the men buy new televisions and chairs, the women buy new clothes. Although the characters are shown struggling to pay bills and to find work, essentially these are comic plot-devices and are not meant to be taken seriously – the friends live wealthy and healthy social and cultural loves, filled with the comforts of middle-class modernity. The New York they dwell in is a very different New York to the one in which most New Yorkers lived at the turn of the century – it is a white-washed New York, where they can hang out without hanging around the diversity and poverty of real New York (again, except for comic effect when Afro-American youth threaten violence). In later seasons, one of the characters, Ross, developed a serious relationship with a black colleague, and the character he dated became a recurring support character until the relationship ended. And there were other one-off or minor roles played by non-white actors. But essentially *Friends* was a mythologized modern America, selling the white American myth of the clumsy white guys winning the hearts of the beautiful white women (Chandler and Monica, and Ross and Rachel). That three of the main characters were Jewish, and one Italian-American, does not alter the fact that the characters were archetypes of modern white American popular culture: the jock, the geeks, the princess, the mother hen and the kooky one. The show (re)presented these archetypes to its American audience, reassuring them that the multiculturalism of society had not gone too far, and showed Western folk and bourgeois classes elsewhere attracted to idea of the West exactly how white people behaved, how they spoke, how they dressed, what things they ate and what they drank: the preppy clothes of the characters, Rachel's hair styles, and the muffin and coffee in the coffee shop became dominant markers of Western civilization before the end of the show.

 Friends and *Star Trek* became global brands in their own right. But American television also spawned reality television programmes, game shows, documentaries and a constant round of comedies and dramas. All of these programmes could be interpreted as perpetuating racial hierarchies and (re)presenting the inevitability and normality of hegemonic whiteness. Historical dramas such as *Mad Men* (set in an ad agency in the 1960s) may be an exercise in myth-making, but they are also nostalgia for white people for a world where the power of white people (white men) was unquestioned.

Whiteness in fantasy films and games

In *Constructing Leisure* (Spracklen, 2011) I discussed leisure activities in the work of Tolkien, and the relationship between his work, his

Englishness and the fantasy genre in books, films and games. I began by pointing out that (ibid, p. 188):

The folk myth of blood and race/nation is easily observed in the work of Tolkien, who was deliberately evoking these notions at a time when England was in the middle of a debate about Englishness (Easthope, 1999). In The Lord of the Rings, the main human character, Aragorn, is a hero and King-in-exile because he has the blood of the higher men, the Numenoreans, running through his veins (Tolkien, 1954/55). His views on taxation, war and peace, governance, diplomacy and so on are irrelevant to his claims to the throne of Gondor: when we read the book, or watch the film, we rarely doubt his pure blooded status, its implications of a supreme race with a holy right to rule the world. Aragorn's claim to the throne is all in the blood – now, of course, we would say it is all in the genes.

Then, after discussing different examples of leisure in Tolkien's books and the growth of the fantasy genre I concluded (ibid, pp. 191–192):

Fantasy fiction had to include elves, or immortal creatures that were bittersweet about the past and had some liminality about themselves, and also orcs, or other-named bad things with a poor education and propensity for violence. Fantasy fiction had to be a struggle between good and evil, and the battle for good had to be won by someone simple-hearted. And fantasy fiction portrayed worlds where ale was quaffed by the flagon, pipes were smoked, and characters spent their evenings telling stories and singing songs in inns that were a pastiche of early twentieth-century English pubs and Chaucerian taverns. The stereotypes of fantasy fiction were established: even those who tried to reject them were influenced by them in rejecting them (Shefrin, 2004; Gray, 2010). The fantasy inn became popularised in late twentieth-century role-playing games such as Dungeons and Dragons and Runequest, which were the pre-internet versions of multi-player fantasy games (Toles-Patkin, 1986). Hence the perfect evening of an Oxford professor became a natural part of the globalised, gaming cyber-culture of World of Warcraft (Chen, 2009), and a teenage boy from South Korea could find it perfectly normal to key in a request for another foaming tankard of ale from the busty barmaid on his computer screen.

Fantasy films like *The Lord of the Rings* have preserved Tolkien's racial essentialization, which is evident throughout his work. Tolkien strongly believed in the essentially good qualities of Englishness. Though not overtly racist, his work stresses the ways in which blood and 'race' combine to create better people: the Numenoreans are the High Men, blessed with strength and intelligence to rule over the lesser peoples. They are white men, and in exile in Arnor and Gondor; they live white, medieval European lives. They are friends with the white men of Rohan, but are threatened and attacked by the evil Easterlings (described as 'swarthy') and the dark-skinned Southrons. The fantasy genre template of races with particular essential biological and cultural characteristics also comes from Tolkien: white-skinned elves representing spiritual goodness and cultural elitism; white-skinned dwarfs representing industry and perseverance; and dark-skinned goblins and orcs representing evil, physicality and base stupidity.

The influence of this racial hierarchy, between different species and within humanity, is evident in fantasy games. Online games such as *World of Warcraft* are, superficially at least, leisure pastimes that are a product of a (post)modern turn. They allow players to create identities, to swap identities, to role-play, to subvert the rules and to engage in acts of intentionality, positionality, hybridity and subversive, resistive agency. *World of Warcraft* transcends national boundaries and social structures. Its reach is global and its players multicultural, but the world it creates is based on a narrow, white, Western medievalism. The mythology of the game is based on the classical fantasy genre created by Tolkien. Different races exist with specific cultures and biological characteristics – elves, dwarfs, men white unless they are different colours (this is always by exception). The publicity material stresses the whiteness of the humans and other 'races' in the game. There is a default northern European/medieval Europe homeworld landscape, and where the adventures move beyond this landscape the exotic Others of Orientalism and imperialism quickly appear. Characters take part in acts of violence and robbery, working together to impose their will on others, ultimately fighting rivals and the forces of evil to become rulers. The game promulgates a Wild West mythology of white Americanism, the sheriff and his men fighting off the Native Americans and the bad guys to win the day. As well as this mythology of white power and the ordinariness of racial hierarchies, the game's myth-making power of intersectionality re-affirms the gender order of hypermasculinity and subordinate femininities; and also the traditional class (white, Western) system of kings, lords, merchants and bourgeoisie at the top, and dirty peasants and ruffians at the bottom.

Another early twenty-first century popular cultural phenomenon is the rise of urban fantasy, and in particularly the dark romance genre of vampire films exemplified by the *Twilight* series. The origin of the vampire story has been well-researched, so I will not dwell too much on it here (see Halberstam, 1993). What is clear is that the archetypal vampire story *Dracula*, written by Bram Stoker, played on contemporary Victorian fears about women's emancipation and sexuality, racial mixing (miscegenation), the physical decline of Western masculinity and the Darwinian terror of blood-sucking Easterners. Most depictions of vampires in the twentieth-century took their visual cue from the chilling, white-faced creature from *Nosferatu* or the aristocratic stage-magician from Universal's *Dracula*. Vampires in twentieth-century films spoke with Eastern European accents but were essentially white aristocrats beguiling naïve young women with their charm and sophistication – Dracula and others like him have always been seen as aspirational characters, villains the audience admired and secretly cheered, even while – and perhaps because – they simultaneously represent the Freudian and racialized Other. When films appeared with black vampires these characters were grotesque stereotypes of black physicality and hypermasculinity (on the film *Blacula*, see Hefner, 2012).

In the 1980s, Stoker's use of the vampire as a metaphor for sex was transformed by a number of egregiously literal authors into stories that used the vampire as a male lover of human women in heterosexual romantic fiction (Burr, 2003). The blood-lust became the lust of teenage girls for the brooding, silent hunk with a tortured past, who only wanted to find his own solace in her white arms. The relationship between white-skinned Angel and blond, white-skinned ex-cheerleader Buffy in the show *Buffy the Vampire Slayer* serves an obvious 1990s precursor to the twentieth-first century evolution of this trope: Angel and Buffy love each other, but when Buffy the human makes love to Angel the vampire with a soul, he loses that soul and becomes the demonic Angelus (Burr, 2003; Wilcox, 2005). The teen angst of Buffy and Angel, and the morality of abstinence, is tempered by the humour, irony and humanity of the rest of the show – though again, there are no black main characters, human or vampire, and the only regular recurring character who is black arrives in the final season as the Principal of the High School (the son of a vampire slayer, who turns out to be on the side of the goodies against the baddies).

In the *Twilight* films, the tone is deadly serious (see Edwards, 2009). A winsome, middle-class white girl, Bella, arrives at her new school in almost complete white town in a middle-class, suburban part of America. Her privileged existence does not stop her from feeling different and alone, excluded from the subcultural rituals of the new school.

However she is soon noticed by, and notices, the tall white boy with the jaw, Edward, who turns out to be a vampire. They find they have much in common and fall in love, but love is not easy, because later on Bella also falls in love with another white boy who falls in love with her, a rebel of sorts called Jacob who turns out to be a werewolf. In the films there is much talk of blood and abstinence. The vampires and the werewolves turn out to be races with different biological and cultural characteristics. The world in which the three of them brood is a very white one. Jacob's surname is Black and it could be argued that the more physical, less cerebral werewolves serves as a metaphor for blackness – the inferior supernaturals losing out to the Aryan vampires. Certainly, in the end, Bella makes Edward her first choice, and after lots of agonizing over miscegenation, and trauma over pregnancy, they have a child together. The films and the books and the spin-offs have led to an entire industry of *Twilight*-lite books, films and television programmes. The vampire has become an unsexed, white teenage boy, who broods for his beloved.

Whiteness, books and magazines

I love books, always have done, and I still think the measure of someone is whether they read or not. There's never been a time since my childhood when I haven't had at least one book on the go. I started out reading fantasy and science fiction but soon moved across to literary novels and classics. For a time I would just buy any Penguin Classic I came across. I am happy reading non-fiction or fiction, pulp or literary, academic or popular. What grabs me first in books is the style, the voice; second, there is the big idea, the argument of an academic text or the meaning of a novel; third, there is the context, the dressing, the imagined place. I'm aware that some people would probably consider me a book snob – I think I am one, too. When I go into someone's house or office I'm always checking out what people are reading, or what they are not reading. Nothing depresses me more than coming across a fellow academic who does not read anything on a regular basis (and there are more of them than I'd like to have believed). I know my tastes in reading are mediated by the publishing industry, by marketing tricks and the invisible mechanisms of criticism. I always judge a book by its cover. I will read reviews of books and buy them – even if I spend most of the time reading book reviews getting angry with the safe, white, middle-class English world in which the reviewer, the author and most of the readers of the newspaper live.

In the Academy there is an established canon of thinkers – philosophers, social scientists, theorists and so on – that underpin

university teaching and academic research. That canon is usually introduced to students in secondary schools, though higher-level critical
thinking and theorists are reserved for university. Different countries,
schools, universities and teachers have different approaches to the
idea of a philosophical canon – some reject the idea entirely, some
favour authors from their own country over foreign authors (especially
where there may be an issue of translation), and of course there is
always an argument over who deserves a place in the philosophical
canon – but there is an essential similarity across the Western world
and elsewhere where Western styles of schooling have a strong influence. In every instance, the canon is based on the sources white, elite
nineteenth-century Europeans and Americans considered to have had
the greatest influence on their civilization. These white Westerners as
I have discussed in an earlier chapter assumed their own civilization
was naturally the best in the world and the source of the best art, culture and philosophy. So the list of books that everybody should read
if they wanted to show they were cultured and civilized was a combination of Classical authors (Plato, Aristotle, Livy, Virgil), early modern
natural philosophers and writers (Descartes, Kant, Shakespeare, Milton)
and nineteenth-century novelists. In the Academy, the philosophical
canon has been transformed by the expansion of critical scholarship
and the specializations that have undermined the idea of the classic,
liberal education – but we are still expected to know about the books of
people such as Darwin, Freud, Kant, Foucault and Marx, and to be able
to refer to them in a way that shows we are confident about what their
big ideas are. This canon of philosophical texts, even with its recent
additions (such as Foucault) is still very much an epistemological construction of white people and an emblem of white power in Western
society. The Culture Wars in the United States were partly an attempt by
radical scholars to undermine that canon of dead white men, but despite
attempts to include ideas and scholars from outside the Western tradition the canon in the Academy remains resolutely white. Furthermore,
this canon is something that has spread across the globe where ever
Western economic and cultural power has spread Western education.

Outside of the Academy, the canon is more fluid, but it still exists as
a marker of cultural capital. To prove one's membership of the liberal
intelligentsia in the West it is necessary to read a number of difficult,
literary works, often novels in translation but always novels in a modern or postmodern style. The older works of the liberal canon are nearly
all products of white, elite Western authors: from Melville's *Moby Dick*
to Joyce's *Ulysses*, the canon takes its older texts from the modernist

literary precedents of contemporary white, Western elite culture. The new canon of the liberal intelligentsia is open to the facts of diaspora and hybridity in late modernity, and there is room in the canon for great works of fiction that explore these facts: there are now a range of authors from a range of countries and backgrounds writing contemporary works about the identity-work of modern life, the tensions of living multiple and fractured lives, and the human condition. Orhan Pamuk, for example, writes postmodern stories about his own country, Turkey, and the tension between East and West, and the modern state and the Ottoman Empire – his novel *Snow* (Pamuk, 2004) explores God, Islam, Islamism and secularism through the experiences of a poet returned from exile to Turkey, who is stranded by snow in a provincial town. In the West, white liberal intellectuals read books like *Snow* to prove they are global citizens and open to the diversity of human experience: but the publishing industry sees only an opportunity to make money out of the exotic, which the white readers eagerly pay for (Coury, 2009). Orhan Pamuk and Arundhati Roy are exotic Others but they are also like the white, Western middle-classes who read their books: the white readers hence get the thrill of the Orient with the reassurance that they do not need to worry about the inequalities of power in history because there are people just like them in Turkey and India.

The other version of the canon is one that has become a global canon of classics, books that have infiltrated popular culture at a global level, though the books were all written in the West. These books are novels specifically written by English, European or Americans authors for popular audiences, or for the elites – but which have stood the test of time to become 'classics', as if accidents of history could be arbiters of aesthetic value. In this popular literary canon can be found: the works of Charles Dickens, Victor Hugo, Jane Austen, Charlotte and Emily Bronte, and many others whose stories have become television dramas, movies and musicals. Shakespeare's plays fall into this category, too. These stories have become part of a collective global heritage, part of a globalized popular culture that owes its existence to the power of the West. The value of the writing is debatable: I have no doubt that Shakespeare was a genius who seemed to be able to balance metaphors and ambiguity like no one else around him; Dickens I consider no better than a hack. But these are the authors who have become the official best of all time, the ones that influence the way people around the world think about fiction, about history, and about humanity. As such they have an enormous distorting influence on what things are considered important in life, and what things are considered to be 'what actually happened' on the past. Most people have heard of Jane Austen, and will think of English stately homes, chaps in white shirts and women in ball gowns – but most people only know Austen from the 'period' genre on screen, not from reading her work. So a particularly elite version of white, Georgian and Victorian England becomes a nostalgic myth for Japanese teenagers: whiteness is normalized for white Westerners watching period dramas and escaping the terror of the modern world; and whiteness is indirectly considered the high point of human civilization and culture.

Bestselling novels are outside of any canon, but have more influence because these are the books people actually read in great numbers. What becomes a best-seller is governed by the persuasive arts of the publishing industry, the opinions of critical reviewers in newspapers and magazines (and increasingly on the Internet), and the fickle purchasing decisions of the book-buying public. A bestseller has nothing to do with the aesthetic value of a book. As I write this book the latest best-seller in the United Kingdom (and one of the top best-sellers in the country of all time) is the egregious *Fifty Shades of Grey*, a badly-written, incoherently plotted novel about a white woman who becomes the sex slave of a posh, white man. There is no redeeming feature about this book – it demeans its female readers, sets back the political struggles over feminism, and is not very accurate about the sexual practices it describes. Despite that, it became a self-published hit that made enough impact to be picked up and sold as a proper book by a big publisher – and now a string of bondage tales are being sold in its wake. Bestsellers do not trouble their readers with difficult questions: they give the reader a sexual thrill or some other emotional hit; and they allow the reader to escape the mundane of the everyday world. As such, many authors and publishers try to write bestsellers aiming for the lowest common denominator of plot and style, and bookshops are filled with crime, thrillers, horror and romances. Women are sold books about women falling in love while juggling modern life, or escapist fantasies about vampires; men are sold books about spies blowing things up and driving fast cars. Some of this no doubt makes for an enjoyable read. The effect of this market-driven production and consumption of popular fiction is the reification of the Gender Order and the white-washing of the modern novel: English or American spies fight strange foreigners from the East; Victorian detectives solve crimes committed by Chinese criminals; anywhere outside of Europe and the States becomes an exotic backdrop filled with colourful locals. Where there is a scene of non-white authors and specialist publishing houses for non-white fiction, such work is often placed in the bookshop under a specialist shelf, away from the other fiction. Men are taught how to be modern men, women are taught how to remain in their place, white folk are told their ways are best and everyone else is reminded of their place in the racial hierarchy. There are authors who subvert these expectations. Much of the horror fiction genre is an ironic attack on the conformity, selfishness and prejudice of the West (usually America), for example, and there are crime novels and vampire novels and romance novels and spy novels and war novels and historical fiction that have non-white characters in central roles. These, though, are exceptions to the norm of white heroes and heroines going about the world being tough men and sassy women.

Popular culture writing is not just carried by books: it is carried by newspapers and magazines, as well, which in turn have been joined (and partly replaced) by the Internet. Newspapers played a historical role in the construction of nation-states and the idea of the public – the section of the populace who were literate and engaged in discussions about politics and culture (Briggs and Burke, 2009; Habermas, 1989[1962]). It could be argued that newspapers have invented the idea of the public, which equates in their eyes to the opinions of their readers, which align with their own opinions (since otherwise, it is suggested by the newspapers, they would not be readers of their newspaper). And it is this conflation of the interests of the newspaper editors and journalists with their readers and the mythological public that has made the particular interests of a section of the white elites in the West become normalized as the public mood, the voice of the moral majority or popular cultural tastes and fashions. In every Western or Westernized country, particular newspapers sell to particular sections of the populace (see Curran and Seaton, 1991, for the British situation). There is generally a number of titles aimed at lower-class readers, what are called tabloids in the United Kingdom, each equally filled with stories about sports, pictures of attractive young women, the affairs of celebrities and 'real-life' stories of great tragedy or joy, along with some basic political reporting skewed left or right as appropriate. Then there are middle-brow newspapers that aspire to capture the mood of the nation, or the middle part of it, filled with right-wing politics of fear and stories of individual success. Finally, there are the higher-status newspapers (broadsheets in the United Kingdom), which aspire to objective reporting of news, critical discussion of politics (leaning left, right and centre) and presentation of the higher forms of cultural capital.

The whiteness of newspapers is clear enough in every form discussed. First, the tabloid newspapers promote a popular culture of consumption, making the norms and values of their (often but not always white) readership seem universal – and those tabloids of the right embrace nationalism and fear of immigration and foreigners. Second, the middle-brow newspapers are even more jingoistic and fearful of outsiders, change and anything remotely liberal or left-wing – these are the newspapers that refuse to accept the modern world and the diverse nature of their countries, which sell to a predominantly older white readership. Finally, the broadsheet newspapers might be above such fear-mongering and hatred, but they still perpetuate the norms and values of the elite white establishment that writes and reads them, whether it is the right-wing white elite (*The Times* in the United Kingdom campaigning for the monarchy and against measures to open up the history curriculum) or the liberal white elite (*The Guardian* in the United Kingdom telling its readers about 'foreign' cinema festivals and other Othered treats to be savoured over a dinner party of lamb tagine). In each case, the readers are told they are the public, and the newspapers present the views of their readers as the views of the people. The Habermasian public sphere has become a cacophony of voices, but the most shrill-sounding are those owned by the white folk in Western countries, the masses and the elites.

Magazines serve more specialized readerships, communities and sub-cultures of interest, like particular websites on the Internet. There are hundreds of magazines that are aimed at non-white readers in America and Europe. However, magazines that are not published by mainstream (national or transnational) publishing companies struggle to get distribution in retail outlets across a country. Magazines that manage to get on the racks at vendors across a Western country are typically aimed at particular niche interests or are magazines for a given gender, or are 'mass market' entertainment. In all these magazines, the images used perpetuate notions of hegemonic whiteness – fashion magazines infamously have struggled to place black models on their front covers; similarly, magazines listing TV programmes or crossword games will nearly always feature someone white on the cover because they are typically bought by older folk (Gill, 2006); even the women in men's magazines and pornographic magazines are white, or non-white models subject to an exotic, erotic gaze (Iqani, 2012). Reading any magazine aimed at 'the general' readership, one realizes that 'the general' very often is another way of saying 'the white'. Popular culture in Western magazines, then, is filled with a hegemonic whiteness that is so visible it disappears out of plain sight.

Conclusions

Popular culture is a key site for the production and maintenance of instrumental whiteness. The hegemony of white, Western culture in globalizing popular culture ensures that diversity, multiculturalism and plurality, while visible in some circumstances, are controlled and limited to spaces and cultural forms that do not serve the interests of the hegemonic white elites.

6
Whiteness and Music

Introduction

I have been going to rock music and heavy metal gigs for more than twenty years, and have been attending folk music and world music gigs for about ten years. I've been a music fan for thirty or so years, ever since I started to jump up and down to Adam Ant with his Ants singing Ant Music on the telly. Like all late modern Westerners I've amassed knowledge of pop and rock music, built up a collection of albums and listened to probably hundreds of thousands of songs by pop/rock artists. Like other white men of a certain age I feel the need to keep listening to rock music and showing off my record collection as a marker of my taste and distinction: this is my cultural capital, paid for over the years and gathering dust on shelves in a number of the rooms of my house. Here are the bands I consider to be classics of their genre (Led Zeppelin, The Cult, The Sisters of Mercy, Iron Maiden, Metallica, Enslaved), here are the bands I want to show to you to demonstrate I am not a follower of fashion, but I am a man of my own individual taste and underground credibility (I'm going to show off and name a few for you, but I won't give away all my secrets, there are some bands I don't want people to know about because then I might stop liking them because they have become too fashionable: Circulus, Mister Fox, Sol Invictus, Behexen, Bretwaldas of Heathen Doom, Nargaroth). I've been to see pop bands, hippy bands, rock gods, new wavers, stadium bands, underground bands, punk bands, goth bands, singer songwriters, indie bands, death metal, black metal, industrial metal, progressive metal, industrial dance, English folk, folk-pop, nu folk, neo-folk, weird folk – and in most of those cases, I've been looking at a stage of white musicians in a crowd that looks entirely white. Where I've been to see reggae, Arabic pop, world roots, the musicians at least have broken the whites-only rule that seems to be hung up somewhere in the concert venues, but even in those cases the audiences seem to have remained predominantly white.

This chapter will focus on whiteness and music. The first section will briefly explore the distinctions between classical and pop, and rock and rap/R&B, to identify the racialized discourses present in music. The next section of the chapter will then proceed to explore world music and roots music, and different notions of whiteness and Otherness present in discussions of authenticity in these genres. The third section of the chapter will introduce new primary research by the author on whiteness and nationalism in English, European and American folk music and European black metal, two forms of music unrelated by sound but with a shared susceptibility to infiltration by far-right nationalists. The concluding section will argue that those who listen to or play music in their leisure lives can make communicative choices that resist instrumental whiteness, but the commodification of most of music industry makes this incredibly difficult and leads to the real danger of unwittingly supporting racialized discourses.

Musical tones: Classical, Pop, Rock, Rap and R&B

Most of the music industry is shaped by Western notions of what constitutes classical music and what is regarded as popular music or folk music (Cook and Pople, 2004). Classical music has been globalized, with strong production and consumption of the musical form in countries and cultures around the world. However, classical music is a product of the West, of the white European hegemony of modernity, and is a continuing legacy of a modernity that shapes bourgeois taste in countries outside of the West – as the adopted taste of Western colonizers and competitors (Wai-Chung, 2004). Classical music is a recent invention of Western European high culture, an evolution of earlier musical forms: courtly music, music for dancing, church music and music played by and for the lower classes. In the eighteenth century and into the nineteenth a number of musicians, composers and elite patrons in Europe worked together to establish spectacles such as the opera, the concert and the recital. Some of these individuals (such as Mozart) were working to challenge bourgeois and elite sensibilities, others were forced to write panegyrics to their masters, while others believed they were introducing music to the wider, urban populaces of European countries. In the history of ideas, the eighteenth century saw the development of a number of theories of aesthetics. Philosophers questioned whether the new music was great art, and if so, why it could not be judged as beautiful or important as paintings or the classical sculptures unearthed in Italy and Greece. Enlightenment philosophers such as Kant tried to argue that aesthetic taste was universal, based on reason and the psychological construction of the mind. In the early nineteenth century the Romantics rejected this appeal to reason and argued that aesthetics were related to something numinous, something felt but not available for inspection by reason. This idea was developed by those who saw great art as a representation of the spirit of civilization, or the spirit of the modern age, or the soul of the nation. A canon of great art, including what we call classical music, was quickly co-opted into an aesthetic of 'race': classical music reflected the numinous, collective spirit of white, Western moderns (either the Germans or the Aryans, or more generally the West). At this period of history, culture was defined as high culture, the culture of the white, European elites. Classical music was taken out with the white elites to its colonies and to cities in other countries where white Europeans held social and economic power, such as Shanghai and Buenos Aires. There are forms of music considered classical in different countries and cultures (for example, Ottoman court music, Chinese opera, Indian classical), and these have a similar level of recognition in their respective countries as Western classical music as being 'proper art' – but it is white classical music (with its orchestras of Western instruments, its conductors, its soloists, its suits and rituals of silent appreciation) that is seen as the norm against which other forms are measured.

If classical music is judged by the hegemonic arbiters of taste to be great art, then the popular and folk music that is defined as being less aesthetically pure than classical music is judged by the white, elite cultural gatekeepers as being anything but art (Frith, 1998). Folk music was initially dismissed as the clumsy music of the lower classes, though since most folk music had become a (re)invention of a national tradition in the last 100 years, folk has taken a more nationalist, racialized position – becoming acceptable as a musical form to the cultural elites, who have gentrified it (Yarwood and Charlton, 2009, and also below). Popular music, pop, is more problematic for the cultural elites. The name itself delineates the music as popular, something for the masses of the modern, urban spaces of the twentieth century, the populace, something that is a product, a business, rather than an art form. Adorno (1947, 1991) was particularly pessimistic about pop music's aesthetic and political potential: he argued that pop music was part of the industry of propaganda and control imposed by the political elites on the working classes to deaden their minds and distract them from the daily acts of oppression, marginalization and alienation. In the first half of the twentieth century, pop music was marginal to the public sphere, ignored by the elites and academics, and viewed as music for the Other: in America, for instance, pop music forms such as jazz and Blues were defined as 'race' music, associated with criminality and disorder (Oliver, 1990); while country music awas for the white poor (Fox, 1992). In the period after the Second World War, pop music started to be consumed by white, Western middle-class teenagers, a generation taking advantage of greater disposable income and greater commodification of Western society to create distinctive, globalized youth subcultures (Frith, 1998). The pop music industry responded by nurturing the growth of pop radio stations then television programmes aimed at young people, cafes with jukeboxes, pop music magazines, record stores and record charts that tracked the sales of singles. When Elvis Presley walked into Sun Records and opened his mouth to sing, the music industry found a white boy who sang with the urgency of a Blues or gospel singer. Elvis became the first global pop star, selling millions of records across the world, not just in the English-speaking West, but in Europe and Asia too (Guralnick, 1995). Pop music and rock and roll followed a simple musical formula of verses and choruses, with simple chord progressions, lasting three minutes: short enough to fit on the side of a single, short enough to be heard on radio. While pop music has been through a number of permutations of genres and scenes, and its industry has suffered from the advent of illegal downloading, it is still a multi-billion dollar global industry built around attractive people singing simple but catchy songs crafted by backroom musicians and producers. Pop stars provide lustful fantasies for young people across the globe; the product has its own regional variations and the best-selling acts follow the formula and embed the whiteness of pop in the paleness of their pop stars (Mayer, 2001).

R&B is a form of pop music that traces its musical forms to gospel, soul and blues. In the United States, this was all perceived (and marketed and marginalized) originally as black music, music made by poor African-Americans for poor African-Americans. R&B (rhythm and Blues) was originally a label used by marketing companies, but has become identified with a particular type of pop music influenced by modern dance music and rap music. The style has supposedly transcended its racialized roots, but it is often problematically used as a catch-all to distinguish pop music that is not (white) rock or mainstream (that is, white) pop music. R&B is still Othered as something 'from the margins', with its highest-profile singers being seen as exotic because of their blackness, or because they are white people singing in a 'black' genre. Rap music has the same problematic racialization at work: it is (re)presented to white music consumers as something authentically dangerous and exotic, a form of music that they can listen to and thrill at the fact that these rough, lower-class (mainly) black people lead troubled lives (Blair, 1993; Yousman, 2003). Rap becomes a way of racializing black men as hyper-sexual, physical, criminal types, Others put in their place by civilized white consumers. Although rappers themselves are involved in complex identity work, and thousands of rappers around the world rap about all kinds of subjects, and thousands of rappers challenge racism and sexism and political oppression (see Lashua, 2007), the rappers that hit the charts (and the rappers who become known to the white people reading the culture pages of their newspapers) are the ones who have become rich selling a racist, white idea of (urban, American) black masculinity (Emerson, 2002).

Rock music has evolved from rock and roll and pop. The genre known to music fans of the twenty-first century has become a form that is supposedly more authentic than pop music – created by bands of musician-artists rather than singers with session players in the background (Frith, 1998). These musician-artists ideally write their own music rather than relying on songwriters and lyricists hired in to make hits (though it is not considered inauthentic to bring in such specialists as producers to 'tighten' the sounds the band records in the studio), and typically rock groups are known more for their albums than their singles. Rock musicians strive to construct a self-image (their own and that of their bands) that conforms to a stereotype of the troubled artist making statements about their individuality and their place in the world: whether it is the individualism of punk and metal, or the romanticism of indie.

The first generation of rock bands used the black music of the Blues as their musicological and psychological template. Myths about Blues singers and their debauched outsider lives were appropriated and (re)invented by the young white Europeans who found global fame as rock stars (Wells, 1983). Led Zeppelin, the band that globalized the rock music paradigm (alongside fellow white Englishmen's the Rolling Stones), typify the appropriation of this make-believe, out-of-control black Bluesman: Robert Plant became the priapic rock god swinging his hips on stage, and his guitarist Jimmy Page dabbled in the Satanism hinted at in the legend of Robert Johnson (who – according to the mythology around him – sold his soul to the Devil in exchange for his ability to make haunting Blues songs). Rock music's first generation copied the Blues in style as well as in identity-making. The sound of rock captured the sound of the Blues in the late 1950s, when it had been elec-trified and commodified in Chicago (Oliver, 1990). When the white rock bands – pushed by major labels and aggressive managers – became global brands, the next generation of rock bands, growing up as white rock fans, assumed rock music to be something created and played by white folks like them. From the 1970s onwards white, male rock fans grew up assuming that rock music was serious and authentic (as opposed to pop and R&B) in part because rock music was played by people they could relate to and wish to be – white men successful in dating women, white men rich on the back of their musical talent. In the twists and turns of indie music and alternative music from the 1980s until the present day, white people's desires to find their selves belonging to some legitimate, authentic musical form has seen rock music become a palimpsest for whiteness. The music industry has played on this racialization by adopting rock music as a white music category, against the 'race' music categories (the industry has not moved far from the days of the early twentieth century, when Blues and gospel were sold as 'race music'). Black rock musicians face endless struggles to be treated seriously as rock musicians, finding their blackness fetishized at best, and at worst their concerts and records failing to sell (Mahon, 2000). White fans and white bands share an assumption about the superiority of rock music against the corporate mainstream, the inauthentic pop industry and the Other: white rock fans do indulge in listening to black music genres (reggae, rap, R&B) as a 'proof' of their commitment to multiculturalism, but black musicians are only okay if they are Othered, the exotic outsiders doing things that are strange to the white rock fans.

World music

World music is part of the wider globalized popular music industry, with its own websites, magazines, record labels, festivals, managers, booking agents and bands. Its definition is, of course, not strictly policed, but it has come to mean 'roots' music from local/national/regional cultures including the West; global fusions and dance music; and pop and rap music from various local/national/regional cultures beyond the West. World music was invented in London in the 1980s as a term to embrace the genres above, while rejecting Western pop music, as well as any form of rock (Bohlman, 2002). The consumption of world music, and the genre itself, is primarily Western in orientation. Musicologists have argued that world music is a product of white, Western hegemony, or more generously, a response to globalization, hybridity and diaspora (for example, Connell and Gibson, 2004). Others such as Corn (2010, on Yothu Yindi) and Skinner (2010, on Amadou and Mariam) argue that some forms of world music are rooted in genuinely local cultures, which serve to define communities and belonging.

It is a cold Yorkshire October afternoon, and I'm outside the Spa Pavilion on Whitby's West Cliff. It is the second day of the Musicport world music festival, and the second year we (my wife and I) have been to it. There is a gathering of people around the vans offering food. All of them are white, English, with middle-class accents. Just like the rest of the people at the festival. Just like us, though I don't think we sound as middle-class as these people sound when we order our organic, lentil burgers.

We want to stand and dance. But it's not so simple. There are two rows of chairs at the back of the hall, for which people have come early to bag for their families. You can't sit in the empty ones because those are seats for the kids who are away at the face-painting stall, or running around the bar area. Some people – families and couples, mainly – have spread themselves out on the floor of the dance area, where they can eat and drink as if they are attending a concert at Glyndebourne. So when we try to stand near the front to dance, we get told off – politely – by the people behind us who are having their picnics. Even when the musicians tell these middle-class white English folk to get dancing, there is resistance from the sitters. As the evening passes by, the acts on stage get funkier, rockier. We try to stand and dance, with some others – we are told to sit down. Before the final act of the night, Natasha Atlas, we are determined to stand and dance. It's Natasha Atlas, Middle Eastern pop via London, we have to dance to it. So we go round the dance area, asking people to join us in standing and dancing. When she comes on, everybody at the front stands and dances, and finally, we can enjoy the festival.

There are tensions at the festival between folkies and poppies; those who want to listen to the music and those who want to dance; and between those who want to sit quietly and those who want to join in. The musicians themselves want to play and be heard, to tell a story and to move people with their art. The people at the festival want to hear something authentic, but there are tensions in what they want, and what they get. Matheson (2008) explores authenticity in the context of Celtic music, and in particular through qualitative research with music tourists attending a Celtic music festival in Scotland. She is interested in investigating the contested meanings and dimensions of Celtic music, and how these relate to questions about the authenticity of the music itself, and the live festival experience. Matheson's respondents identify something authentic in the music and the experience of listening to it at the festival. For some, there is something about the way the music is performed that is essentially Celtic: particular instruments such as bagpipes or tin whistles, or emotional vocal performances. This leads to complaints about the presence of more commercial acts at the festival, whose Celticness is not so obvious in their music, and accusations from some respondents that the festival had become inauthentic through this acceptance of innovation and commercialized taste. For other respondents, the festival serves to authenticate their own Celticness, through their emotional response to the music. One respondent argues that to feel moved by the music, to become emotionally involved in it, listeners need to have 'some Celtic blood' (cited in Matheson, 2008, p. 69). For others, the emotional response to the music was a way of becoming Celtic: establishing their own personal identity as a spiritual successor of the Celts, a member of the imaginary community. This, Matheson argues is the central theme in her research: 'the emotional aspect has a role to play in the ways that event consumers interpret the authenticity of the music … it signifies that theoretical readings of authenticity demand a closer examination of consumer's emotional attachment and reactions to cultural rituals and experiences' (Matheson, 2008, p. 70).

There's an acoustic, roots band with two members – one from Estonia, one from Wales. They combine the traditions of the folk music of both countries, with some original compositions and fusions. Only a handful of people are here to watch them, unlike the bands from Africa or India. I wonder if perhaps this band is just not exotic enough. There are no stereotypes here of vibrancy, no perceived purity of culture and tradition, no make-believe authentic nativeness, just some nice tunes. I buy their CD – there is no queue to get one, no scrum of people telling them how great they are to come over here to play for us.

At the festival, all authenticity is bracketed by the need for musicians to sell their products; for the market traders on the stalls to shift hats, clothes and 'ethnic' jewellery; and for the burger van to sell enough to make a living. For the consumers at Musicport, the more exotic the world music, the more authentic it is. When one Middle Eastern player makes nationalist comments, there is polite applause: here is someone who is an authentic voice of the oppressed, a true Other who sees the world through the lens of an acceptable chauvinism.

On the Sound of the World forum (http://www.charliegillett.com/bb/), there is little in the way of critical debate about the nature of world music. The authenticity of the product is assumed to be uncontroversial – if it has come from abroad, and it is not a form of rock music, it is world music. Anything that fits the definition of world music given above is deemed to be authentic – both of the genre and of the national culture out of which the musicians have emerged. There is a simplistic belief in the moral goodness of world music as being something truly representative of something counter-hegemonic, contrasted with a Western pop music industry that is derided. So, for instance, poster Adam Blake had this to say in a discussion about the decline of the music business (posted 21 July 2010, 12.17 pm, in the thread 'Andy Gill: Music Business Dying', on the Sound of the World Forum):

> I have very little sympathy, if any, for the pain of the music business. They brought it on themselves. I caught the fag end of it in the late 80s when I was briefly a journalist writing for Music Week – the trade paper. The amount of WASTED MONEY – it used to beggar belief. They threw it all away. The hubris, the arrogance, the endless bullshit, folly and foolishness. Older lags than me will say, 'ah, but it was such fun' and I daresay it was. Well, here is the harvest that you reaped, you stupid nincompoops. Enjoy it, breathe it in. If you got rich while you could and you didn't put it all up your nose, good luck to you. The rest of us will be down at Cafe Oto, or Passing Clouds or someplace similar, checking out some music, maybe buying a hand-made CD straight from the band if we like what we hear.

The post is typical of its kind, and demonstrates the way in which world music fans on the forum distance themselves from the distasteful commercialism of the rest of the commodified music industry – while trumpeting the more authentic experience of listening to live world music at some modish venue and buying a *hand-made* CD. It is almost as if the world music scene is something that has been created *sui generis*, free from any taint of capitalism or factory-based technologies. Of course, this is just a fallacy – the CDs may well be hand-made (copied direct from a computer to a CD-ROM), but the musicians, engineers, producers, publicists, managers and roadies operate in an industrial complex where the artistic end product only exists at the sharp-end of a series of instrumental compromises. There is a Goffmanesque play of authentic, exotic Other that hides, like the curtains in the Palace of Oz, the old man pulling the levers.

On the fRoots forum (http://froots.net/phpBB2/), Ian Anderson, the editor and founder of the magazine, had much to say about world music in a thread started to discuss the demise of an American reggae music magazine ('The Beat RIP: world music and its lost soul', in The fRoots Letters Board, response by Ian posted 22 December 2009, 2.09 pm). His anger with the current scene was clear, and it is worth quoting at length from his posting:

> As the world music scene has tried ever harder to become glossy, coffee table, and ape the corporate mainstream with its Thatcherite 'fuck

you, I'm only in it for myself and what I can get out of it' mentality, it has lost a lot of its soul in the process and...its sense of community and the all important symbiosis that used to make it thrive...I had a lightbulb moment recently. The generation largely in charge of the world music scene, regardless of their politics, really *are* Thatcher's children. They grew up – spent their teenage years – in the 1980s and those attitudes, subtly or unsubtly, infected everybody in their formative years. Whereas the folk scene had a lost generation and it coincides with that era: with exceptions of course, most folk scene people seem to be either well under 30 or well over 50. Coincidence? Wild theorising and clutching at straws? Dunno...Once upon a time when the scene was young and there was only a relative handful of movershakers at large, world music magazines, radio DJs, labels and intrepid promoters used to co-operate, communicate with and support each other out of general enthusiasm, characterised by the 1987 campaign that named the box. Very little now. This is thrown into sharp relief by the UK folk scene which is thriving, expand-ing by leaps and bounds, and has always been powered by the very grass roots mutual support that the world music 'industry' has largely lost. I've lost count of the world music labels who 'forget' to send us review CDs or tour dates of artists who we've been among the first to enthuse about, sometimes given cover features to in their beginning days. I'm bored with hearing that 'we don't need to advertise because you're going to write about it anyway.' We really notice simple things like the way that UK folk artists and their labels quite regularly say 'thank you' for features and reviews, whereas the world music equiv-alents barely ever bother to even acknowledge them. Except when they want something else for free the next time. Very 'children of Thatcher', the Tottenham Court Rd shop assistant syndrome. That's one of the lesser reasons why I didn't go to Womex this year, for the first time ever. One of the many things which crystalised this decision was that in 2008 we took along a staff member who was, shall we say, a little arsier than the rest of us. While the usual endless streams of people were coming up to our stand pitching for features and reviews, she was asking sweetly whether they were a subscriber or advertiser. Almost all made it clear that they weren't and had little intention to be. Except for those who assumed they could buy 'advertorial' space, as in certain other mags, and were quite shocked that they couldn't.

Anderson's comments are interesting because of his key role in the scene as an editor and publisher. His magazine supports world music artists, but he is more ambivalent about the world music scene's authenticity.

It is not as authentically communicative as the folk music scene, which it overlaps. Anderson believes that the latter music genre has some connection with traditions, with roots, and with communicative actions, which make the creation and reception of folk music less instrumental than other forms of pop music. Some world music falls into this authentic folk/roots scene by nature of its relationship to 'genuine' traditions. Anderson's post attracts some criticism from a small number of posters who argue that world music does have decent people who are motivated by the music and not making money working in it. However, most of the posters agree with Anderson's suspicion of world music as something instrumentally ersatz. One poster, Vic Smith, makes two important contributions to the debate. In his first post, he reflects on his experience of world music gigs:

> I like the majority of the people that I meet there, but there is always an element that are there to show how 'right on' they are, giving off that 'Here's me doing the right thing politically. I don't give a monkey's about the music or the tradition it comes from.'
>
> (posted 22 December 2009, 11.37 pm)

Two days later, he continues Anderson's critique:

> The world music scene did reflect a growing interest in the roots music of different cultures in the western world, but it was always a more disparate movement whose very existence was codified at that famous meeting some 20 years ago. From its inception, 'world music' was partly a marketing term to give a place to the records and magazines that were breaking out to reflect this interest. By its very nature there were bound to be huge differences geographically as well as culturally. The possibilities of world music being exploited by unscrupulous record producers, agents, managers etc. were enormous right from the start.
>
> (posted 24 December 2009, 9.32 pm)

Across the forum there is a dominant discourse of authenticity in folk music, which is identified with communicative actions around the meaning and purpose of musical creations, and the active engagement of fans in the establishment of folk/roots taste (Bourdieu, 1986). It is this discourse that contributes to the maintaining of the boundary between true 'roots' music (folk and world) and pop. There is, politically, liberal-left scepticism of global capitalism, a desire to preserve English folk traditions but a strong rejection of English nationalism. Support of English 'trad' folk musicians is given alongside similar acclamations of support for world musicians such as the above-mentioned Yothu Yindi, who can demonstrate their commitment to preserving their own musical traditions and their own 'authentic voices'. That said, the posters on the fRoots forum are also supportive of English folk musicians who play with the traditions of the scene, or who embrace hybridity to create new fusion forms of roots music: Bellowhead, who incorporate jazz and Latin American rhythms in their music, are praised for their folk grounding and their commitment to re-working traditional English folk songs; and the Imagined Village project, a multicultural clash of traditional English folk with British Asian and black British music and musicians, is similarly praised for its grounding of folk in modern Britain's diasporic hybridity (the band's name cleverly adopting the academic framework of the imagined community).

Folk and metal

I have published elsewhere on the similarities between English folk music and black metal's use in the construction of hegemonic, instrumental whiteness (Spracklen, 2012). This section of this chapter develops and extends that analysis to discuss folk music in general, and the wider heavy metal scene. I discussed English folk music and black metal in my previous research because they were musical scenes I was involved with, and I knew from my knowledge of some of the black metal bands that there was a conscious effort by English black metal fans and musicians to become interested in English folk. There was also evidence that both music scenes were attracting the interest of white nationalists and other far-right activists: the British National party (BNP) in England sold English folk music through its website and its leader listed various English folk acts as ones he listened to; and white nationalists were involved in the production and sale of a number of black metal recordings. Black metal is a form of extreme metal associated with elite ideologies (of purity, of misanthropy, of Satanism, of heathenism and nationalism – see Lucas, Deeks and Spracklen, 2011; Spracklen, 2006, 2009). Despite its name it is white metal, metal initially created by white musicians in Northern Europe reacting against artificiality and trends in the metal scene, and attempting to preserve metal as a true, underground, anti-modern, Satanic movement. Black metal of course has globalized and become an established genre of heavy metal, but its musicians and fans still argue over its status as outsider music: underground, elitist, communicative and non-commercial. Heavy metal is a form of rock music, which has become an established genre of its own since the 1970s. Heavy metal takes to excessive levels the amplification, individualism, pompousness, hypermasculinity and theatre of rock music. In the history of heavy metal most of the leading bands and musicians are white Americans or white Europeans, white men singing songs about power and freedom and individuality that have become a part of the globalized, pop music scene. Heavy metal itself has become globalized, with bands emerging in many Westernized countries, or in Westernized classes in countries beyond the West. In the West, the music has become so popular and homogenized that it attracts young fans of all classes and from all social groups. Despite this construction of heavy metal as the pop music of every person wanting to express their true individuality, heavy metal remains in the West a strongly white musical subculture – a music for white trash, played by white men with long hair and beards, listened to by white folk who want to be associated with its faux-outlaw status. Metal is a significant part of popular culture in the predominantly white Scandinavian countries and in the countries of eastern Europe: in the former, metal becomes a way for white teenagers to associate with an invented white Viking past; in the latter, metal becomes a way of expressing national, white purity. It is sold by the industry as white music, and black musicians playing metal have to work hard to be accepted as 'true' metallers.

English black metal bands have taken an interest in English folk music because they are seeking to find supposedly pure roots in an imagined England of 'yesteryear' (Spracklen, 2012): the themes of their songs are about an older England, and folk music offers another way of playing with those themes. As I have previously argued, English folk music is an invention, a construction of a romanticized, mythical time and space where England's rural folk lived ideal lives in tune with the community and the land. Although many modern English folk musicians are aware of the make-believe mythology about Englishness in the folk tradition (and many of them play with those traditions), English folk is defined by its status as something supposedly more authentically 'English' than pop or rock.

Folk and roots music worldwide serve a similar nationalist function. In the United States, old-time music preserves the simplicity of a rural America populated by recent white settlers, one where 'heroic' white Europeans won land and farms from the wild country. It is an America untroubled by urbanization and mass immigration, one where the racialized politics of the recent past is elided into the smoke of wood-burning stoves and the birdsong of mountain forests. Old-time music is a modern construction, like country music, which was created in the twentieth century by marketing offices and music retailers to distinguish white music from the black music of the Blues (Oliver, 1990). Old-time music is a product of middle-class white musicians and musicologists trying to find the authentic roots of country and rock music – some trace of an older, purer life in some old white village high up in the mountains. The fact that old-time, country, gospel and the Blues interacted and influenced each other for the first half of the twentieth century is ignored by those who want to find the most authentic (and hence the most 'white') source.

European folk music is similarly tainted by the search for authenticity, national identity and white purity. In the revival of the various Celtic music scenes, the racial purity of the nations is hinted at in the call for preservation and cultural pride: the Bretons demand more support for Breton music at the same time as demanding rights of national and cultural identity (Gemie, 2005). But such cultural identity is given prominence in the region at the expense of other cultural identities that might exist in the same locality, so Breton folk music easily slips into becoming a site or exclusion and promoting whiteness against recent migrants into France. Other forms of folk music operate at a national level. In Hungary, folk music becomes a way of constructing an exclusive form of nationalism that excludes minority communities and supports greater Hungarian jingoism (Frigyesi, 1994). In many other countries, folk music is played by nationalist parties at rallies, where it becomes a way of associating the nation with a particular history of a particular people – the music is racialized as the music of the dominant 'race', which in Europe means folk music becomes associated with whiteness. Nationalist folk music exists in other countries outside the hegemonically white imaginary community of the West, of course, and serves a similar function of creating and reproducing exclusive forms of national or regional belonging. In the West, where this music is often consumed by white fans of world music, it becomes known as roots music. As I have already discussed, white, Western music fans and music companies seek out these supposedly authentic forms of roots music and consume them. Whiteness in folk music, then, is constructed in the act of creation and in the act of consumption. In the act of creation, musicians look to fix their creativity in the narrow bounds of a particular tradition, which in most Western forms of folk music is an imagined, white past where white folks constituted the population of the given nation or region without any troublesome migrants or minority groups confusing the pure fantasy (the actual facts of migration and heterogeneity of any given space are always put to one side in the fantastic imagination). In the act of consumption, white music fans find either a reconstruction of their own white pasts or some (post)colonial exotic interaction: they search for roots and find the music of their 'folk', the nation or region with which they identify most strongly, either through accepting the music of their tradition (and the make-believe connection with the past) or – ironically – in the appropriation of another folk music that is strangely Othered enough to confirm their white self-identity in the shadow of its consumption.

The similarities in the creation and consumption of the music in the folk and metal scenes should be reasonably clear. As I conclude about English folk and black metal (Spracklen, 2012, p. 12):

> Both music scenes can be understood to represent places where individual actors have space and freedom to resist commodification and to insist on their own arbiters of taste, of belonging and of exclusion (Prior, 2011). However, both music scenes are reliant on the machinery of capitalism and the discourse of industrial production: metal and folk musicians make records and companies sell those records to fans who find their identity through the transactions of consumption. Such freedoms are also compromised by the hegemonic whiteness of both music scenes. Whiteness is tied up with elitism in black metal and Englishness in folk music, and both forms of whiteness make these scenes attractive to fascists and racists seeking to co-opt them. While English folk music has been quick to resist fascism co-opting its traditions, black metal is open to fascist involvement through its adherence to individualism. In both scenes, however, hybridity is rejected in favour of purity (of ideas, of traditions, of symbols of belonging and exclusion). Purity creates impermeable boundaries, which dissuade taste-makers and scene-setters from embracing hybrid forms, typified by mainstream heavy metal and the globalised pop of world music. Hybridity here is key to a critical understanding of the role of music in the construction of multiple identities, but such construction, while demonstrating the agency presupposed by Brah (1996), Solomos (1998) and others, is limited by the instrumentalized structures of Western society and the whiteness of Western national identities (Garner, 2006). This leaves black metal's true, 'kult' nature as elitist and nihilist, with the white face make-up of the corpsepaint a suggestion of the white mask of Fanon, the whiteness of the scene's mythology and history; and

leaves English folk music suffering the dissonance of rejecting fascism while policing the boundaries of a mythical English village in which white people dance around a maypole.

Conclusions

The music scene in the twenty-first century exhibits all the symptoms of late modernity. We can see here how the work of Habermas can help us make sense of music as leisure, or musical leisure. Musical leisure is in late modernity a fundamentally commercialized activity. It is, to use Habermas' terms, dominated by instrumental rationality. This does mean there is no space in musical leisure for agency and resistance – there is still some communicative space where individuals can use music to make meaning of their lives, make connections with others, make choices, make identities, counter instrumentality and resist hegemony. But as the century progresses those free, communicative spaces become fewer. Music has been commodified, packaged and sold in a globalized music industry. At the same time, musical tastes and fashions have become an important marker of distinction and cultural capital. In the West and beyond, those who buy the dream of individualism seek authenticity and belonging by making or listening to a range of music genres. For those seeking elite respectability and distinction, classical music concerts provide a space where they can prove their cultural superiority over the masses who listen to pop music on the radio. For those wishing to demonstrate their taste and resistance to marketing, there are world music and folk music forms to use. Pop and rock music collectively sell individuality, conformity, Western idea of true love, hegemonic masculinity and subordinate femininity, belonging to a scene or a genre, a stake in modernity and a commodified way of life. There is some communicative, creative activity in musical leisure, the creation of music for the sake of creating music and the sharing of creation with other musicians. But most musical leisure activity does not involve the actual construction of music, alone or shared. In fact, most musical leisure activity is in the consumption of music as a music fan: listening to recordings, using electronic equipment and digital media, discussing music with other fans, buying products associated with particular scenes and bands, watching gigs.

The whiteness of pop, rock, classical and many forms of folk music has been demonstrated in this chapter. Is this whiteness hegemonic – is it instrumental, or working to preserve instrumental hegemony? I think this can also be demonstrated. First of all, there is an invisibility about the ways in which such whiteness operates: it is not just the ways in which folk and metal are sold as white music, but the ways in which rock and pop and classical are seen as 'normal' forms of music against which the exotic Others of rap/R&B and world are constructed. In this invisible racialization process, there is a massive disparity of power and agency. Musicians and the music industry in the West retain the hegemonic power of the West to dictate the shape of global economics and global popular culture: white forms of music are seen as the golden standards of modernity, promoting individuality and belonging and exclusion. It is perhaps obvious that modern classical music is a Trojan horse for hegemonic whiteness, with its explicit links to European debates about high culture and the superiority of white, Western traditions. Pop/rock music and folk music might be less obvious as carriers of hegemonic whiteness. It could be argued that both music scenes are counter-hegemonic or residual in cultural terms, as they are genres derided as inferior by the tastemakers of the elite classes. But both genres are put to use in the construction of hegemonic whiteness. In Western folk music, a tradition of purity and exclusion is created through the nostalgic (re)production of a whiter past – the founders of many folk traditions in the West were drawn from the white ruling classes and desired to use folk music to justify nationalism and exclusionary ideologies. In pop/rock music, the tight rules of the industry have become globalized but the scenes are still places where whiteness is defined through the dominance of white musicians and industry workers, and the use of racialized marketing tools. Pop music may be an instrumental tool of oppression, but it is also one that reaffirms the racialized politics of twentieth-century America. Those who listen to or play music in their leisure lives can make communicative choices that resist instrumental whiteness, but the commodification of most of music industry makes this incredibly difficult and leads to the real danger of unwittingly supporting racialized discourses.

7
Whiteness and Sport

Introduction

Here I am as a young white boy, playing rugby and soccer, and running and cricket, not very good at any of them but doing it all like it's the natural thing to do. We play sports in our own way, with our own rules – these are informal sports practices, more like other games we play than the grown-up versions our teachers and fathers talk about. In playing soccer I learn the importance of competition – I need to win, to beat others, whatever it takes, even in the bounds of the old bowling green we play on. My friends feel the same. We are boys becoming men, learning to be men, to be tough, to be ruthless in the quest for glory, watched by the girls on the benches at the edge of the green. In playing rugby I learn to keep my tears to myself, to smile and joke at the cuts and bruises and sprains. My brother, another white boy learning to be a proper white Yorkshire man, breaks his ankle at rugby training and walks home limping. We all soak up the lessons of football and rugby: sports for men invented by white, British men, our sports that had spread with our British customs across the world. We understand the importance of England's success in the World Cup and other test matches, and dream long, crazy dreams of pulling on white jerseys and playing for our country. As I get older I become more of a spectator, learning to get my sports fix like other young white men through the transference of my energies and desires onto the mainly white men in the teams I support. I notice that there are black men in those teams, and I think of them as exotic sports superstars, the West Yorkshire equivalent of Muhammad Ali. I know there are non-white people who live in this area, and know some of them as friends (two of them, to be precise), but I don't consider them to be strangely exotic. They are just friends of mine. The black footballers and rugby players are aliens to me – and while I am drawn to their strangeness I find it hard to see that they are the same as my friends.

It's only much later – when I start thinking about these things in a critical
way, when I start becoming politically active in my teens – that I realize the
exoticness of the black players was caused by my younger self's certainty that
the sports I followed were 'normally' white (British), a representation of some
make-believe white Britishness associated with the mythicized England of Hur-
ricanes and Spitfires and empires (I deliberately conflate Britain and England
because that is the way this nostalgic, white nationalism works, mixing up
the two categories without any care to give each country its correct place in
the convoluted, post-colonial but still faintly feudal United Kingdom of Great
Britain and Northern Ireland).

This chapter will focus on whiteness and sport, and will draw on
secondary analysis of sociology of sport research, along with some primary research on the history of modern sports and contemporary sports.
The chapter will draw on examples from the United States, Australia,
South Africa and Europe. The first section of the chapter will examine sports participation and the construction of cultural capital through
involvement in sport. Whiteness will be identified as an invisible, taken-
for-granted signifier in modern sports, with participation among white
people in certain sports such as basketball and athletics being dependent
on sports that provide safe 'white spaces'. When significant numbers
of black people start to take an interest in these sports, I will demonstrate that white people start to choose other sports to play. The second
section of the chapter will focus on sports fandom and the modern,
professional sports industry. I will show that white people find an imagined community of 'pure' whiteness in supporting a particular club or
national sports team (for example, the South African Springboks) and
this whiteness is only partially challenged by the introduction of black
athletes into those teams. White sports fans, I will argue, are (generally,
mostly) comfortable with black athletes representing their club or their
sport or their country because the globalization and professionalization
of sport has turned young black athletes into caricatures of physicality
and modern-day slaves. White sports administrators are still in charge
of the sports, still defending the white history of these sports even as
some black athletes are embraced as exotic 'ringers'. As such, the power
dynamics ensure whiteness remains hegemonic.

Sports participation

Modern sports are an invention of hegemonic, Western instrumental
whiteness. Of course, there are sports-like activities that owe their origins in different cultures and societies, such as Turkish wrestling and

the variations on polo still played in the highland zones of Pakistan and Afghanistan (Parlebas, 2003); and there are sports-like games that appear in the historical and archaeological records for Central and South America (see discussion in Spracklen, 2011). Even in the West, there are precursors to modern sports, especially in the period of the Classical Age of Europe, when Greeks and Romans played and watched a number of games that are still part of the modern sporting arena (athletics, gymnastics, wrestling) as well as games that are no longer considered to be sports at all (chariot-racing, gladiatorial combat). Despite those exceptions and caveats, the first sentence of this paragraph still holds true: what we think of as sports, the things we play and the things we watch are all inventions of white, Western modernity.

The idea of sports first appears in the Early Modern period, when European elite men used the phrase to describe their leisured, gentlemanly pursuits in the outdoors: hunting, shooting, fishing and horse-riding (Borsay, 2005). At the same time, rural games such as football were identified as the sports of the poor: these games were described by gentleman authors such as Strutt, who believed they were the remnants of a residual, feudal culture, with informal rules and governance (Strutt, 1801). In the early nineteenth century, the white, Western elite classes started to take an interest in sporting activities that were threatened by the rise of urbanization and industrialization. Educators and Christian activists believed that the old, folk sports were a valuable learning tool for young elite men: they should take part in sports to be physically fit to run the empires of the West, and morally fit to be good Christians. Sports would provide these young men with the bodies and minds they needed to keep Western, hegemonic whiteness in its hegemonic state. As the working classes moved to live in towns and stopped playing sports associated with rural, feudal cultures, these same sports were adopted and codified by the elites in the schools and other institutions – whether it was soccer in the British universities or baseball (adapted from rounders) in the United States. These sports were believed to be expressions of white, Western modernity, of a shared elite civilization – even where nationalism was also a factor of the modern sport's growth (again, baseball is a classic example). As well as seeking to use the older games as the building blocks of carefully regulated and governed modern sports, the Muscular Christian movement adopted and adapted games from the Classical Age, replicating the similar neo-classicism emerging in elite culture, architecture and literature.

By the second half of the nineteenth century, governing bodies of sports had emerged throughout Europe and North America, organizations with individual or club members, which decided on rule-books, competitions and laws over who was allowed to take part in the particular sport. This was the birth of modern sports – sports could not be modern without fixing their rules, developing governance structures and becoming part of the regulatory regime of modernity (Spracklen, 2011). Some of these governing bodies had a domestic remit; others were international organizations that had domestic governing bodies of sport as their members. These governing bodies were initially dominated by the elite classes of their countries, the white Westerners – so, for example, when soccer developed in Argentina it was dominated and controlled by British men (Archetti, 1995). Modern sports' governing bodies became quickly fixated on the fear of professionals, lower-class players being paid to play as opposed to middle- and upper-class gentlemen amateurs. Rules on amateurism tore some sports apart, such as rugby (Collins, 1999; Dunning and Sheard, 2005); some elites lost the battle and professionalism emerged as the legitimate (and dominant) practice in some sports (for example, soccer); but in most sports, amateurism survived as the prevailing ethic into the second half of the twentieth century. This ideal of the amateur helped to preserve the whiteness and the maleness of sports participation: amateur sports such as athletics, rowing, American football and rugby union became the sports played by the white, Western elites in their private schools and top universities. Professional sports such as soccer and baseball were used to keep the poorer classes (white and black) in their place in the Western, modern system, through offering poor men dreams of becoming famous, a chance to cheer their local side and a chance to bet and drink with their friends.

There were, then, two routes to sports participation, which both excluded marginalized groups such as women and minority ethnic groups to a greater or lesser extent. For the white elite men, amateur sports were something they did to become proper men, physically and mentally equipped to rule and give orders, fighting in imperial struggles and capitalist marketplaces. Amateur sports could be obscure such as real tennis or the Eton Fives game, or they could be sports that were well-known and played by a wider range of classes including the bourgeoisie (such as golf). Especially popular were team sports such as football and basketball: these sports promoted the importance of victory, individual endeavour and physical prowess, alongside the importance of working as part of a team to bring glory to one's club. Sports such as these became fundamental in schools, colleges and universities, where the new white elites and the growing white middle classes of the West were trained. As the twentieth century progressed the importance of sports in the elite education curriculum was established, normalized and allowed to spread to women's educational establishments, and throughout the world where modern, Western values were adopted. The second route to sports participation was the evangelization of the working classes by sports advocates, men (and occasionally women) trained in the elite system determined to make the poor better in body and soul through sports. This led to widespread adoption of formal physical education curricula and the rise of State intervention in creating sports facilities, as well as the increase in professional sports activity when young, working-class men realized they could make money from playing sport. This second route to sports participation did allow some non-white participation in sport in Western countries, but the participation was controlled and shaped by the condescending attitudes of the spots advocates. Sports were still viewed as something done at an amateur level by rich white men – sports for poor people were ways of making them healthy, pliant and less aggressive, a way of giving them something to forget their troubles.

Modern sports were fundamental to the resurrection of the Olympic Games as a neo-Hellenic, elite, amateur activity. This resurrection was linked explicitly to the survival of the hegemonic, instrumental whiteness of the Western elites. In 1908, reflecting on the increasing importance of the Modern Olympic Games in the wider geo-political landscape, their founder Baron Pierre de Coubertin expressed his hope that the Games were having 'an influence, which shall make the means of bringing to perfection the strong and hopeful youth of our white race' (De Coubertin, cited in Carrington, 2004b, p. 81). For De Coubertin, the Modern Olympics were beyond nationalism, but not beyond the symbolic boundaries of early twentieth century notions of race. The Games, as King (2007) has argued, were designed and promoted by an elite section of white European society at the end of the nineteenth century as a means of preserving and promoting elitist ideas of belonging and exclusion. The rhetoric of the open playing field symbolized by the Olympic rings masked the reality of sport's role as the maker (and marker) of racial difference (Mangan, 1981, 1985, 1995; Mangan and Ritchie, 2005): sport made white men fit to serve the engines of commerce and empires. The ideology and ethics of the International Olympic Committee (IOC) remained strongly shaped by its founding father's beliefs. The IOC was dominated in the first half of the twentieth century by Europeans, Westerners, white men trained in the national governing bodies of sport of their own countries. These white men tried to keep the Olympics free from 'politics', which for them was away of preserving the power of the white, Western hegemony over the rising economies and societies of the rest of the world. They defended amateurism and rules of governance designed in the nineteenth century by European sports governing bodies, and supported the use of sport to protect the interests of the ruling classes. At the 1936 Berlin Olympics, the IOC's white men found a common purpose with the Nazis, adopting the torch race staged by the Germans for future events and praising the Nazis for their determination to preserve 'European' civilization (Keys, 2004). The political struggle in the Olympics movement started after the Second World War, when countries freed from imperial control pushed for greater democracy in the IOC, and for sanctions and boycotts against openly racist regimes such as that of South Africa. But although the struggles brought changes, and the IOC embraced professionalism, the Olympics still remained a Trojan horse for white, Western values and hegemonic power.

In parts of the West where there was nationalist struggle, sports became a way of identifying with either the local elite or the oppressed cultural group: in Ireland, the Gaelic Athletic Association (GAA) allowed Irish nationalists to resist British imperialism through supporting hurling and the Irish version of football (Fulton and Bairner, 2007). In parts of the West where institutional racism was a part of everyday life, such as the United States or Australia, sports were racially segregated through official sanction, and the white-only governing bodies were favoured when it came to money and ownership of sports resources (pitches, fields, equipment). In the United States, so-called negro leagues and colored clubs were run in baseball from the end of the nineteenth century to the second half of the last century: these were partly acts of resistance and autonomy, but also acts of control and subservience – a way to keep mainstream baseball all-American white (Lomax, 1998). In Australia, the dominant football codes and many other sports operated unwritten colour codes to bar the involvement of Aboriginal players (Tatz, 2009). In India, the British sport of cricket (which had managed in its birthplace to allow professionalism for the lower classes while allowing the white, British elite classes to retain power over the game through its arcane organizations) was the site of racial tension and white privilege – it was seen as the sport played by the colonialists and their collaborators in the Indian elite, though it soon acquired a following among the wider population. A similar state of racial segregation operated in cricket in the West Indies and Australia, where formal and informal practices of racism ensured the whiteness of elite clubs and governing bodies well into the twentieth century.

South Africa provided the most egregiously overt form of racial seg-regation and instrumental whiteness through the institution of the Apartheid regime. The white Souith African elite were strongly influ-enced by racist ideologies and notions of white biological superiority that were part of the popular discourse of the twentieth century. This belief in the purity and superiority of the white race(s) was central to the Nazi ideology of Germany, which led to the horrors of the Holo-caust. Despite the lack of scientific evidence for white racial superiority and purity, and despite the dire consequences of following that logic in Germany, such ideology continued to be dominant in the West after the Second World War. As Western empires retracted, white people clung to these beliefs, especially so in South Africa. Putting into action across the entire country policies and practices that operated locally in many towns and districts, the system of Apartheid attempted to per-manently protect the interests of the white ruling classes against the non-white majority. This was racial segregation as seen in other coun-tries such as the United States, but on a much larger scale, and with national legislative power. White and non-white were given separate spaces, buildings, services, goods and activities – with the white pop-ulation having the best share of the resources and the power, and the non-white only allowed access to the worst part. Sports played a cen-tral role in maintaining Apartheid and the racist power relationships: white South African national teams in cricket and rugby union were celebrated abroad, touring teams visiting South Africa observed and sup-ported racial segregation (and often failed to bring non-white players), and even when formal boycotts were established white European, white Australians and white New Zealanders continued to find ways to visit the country and play sport with the white South Africans (Booth, 2003).

In the second half of the twentieth century significant campaigns for civil rights and racial equality changed the legislative and cultural landscape of the West. This occurred at the same time as increased postcolonial migration into the West, along with the emergence of non-white middle classes. On the one hand, Western nations tightened their border controls and joined in rhetorical posturing and racialized fear about foreigners swamping the national culture – a rhetoric that con-tinues to the present day in most Western countries. On the other hand, governments encouraged organizations to promote their services and activities equally across ethnic groups, and also built into leg-islation commitments to promoting diversity. Even South Africa was forced to abandon its segregation after the fall of the Apartheid regime and the birth of the 'Rainbow Nation' under the governance of the African National Congress. The fact of multiculturalism could not be

denied in the West, even though considerable numbers of white people, of all classes, resisted it. In sports, this meant an opening up of
recruitment and development policies, the establishment of strategies
targeting minority ethnic groups – and the increased involvement of
minority ethnic people in various sports (Long and Spracklen, 2010).

But there is still today a construction of white, elite and bourgeois
cultural capital through involvement in sport. Just as it was present in
sports participation in the history of modern sport, whiteness can be
seen as an invisible, taken-for-granted signifier in modern sports, with
participation among white people in certain sports being dependent on
such sports providing safe 'white spaces'. In Great Britain, white people dominate participation in sports associated with wealth and time
resource, the two things elite classes have access to: so sports such as
rowing, cycling, equestrianism, swimming and sailing (here the categories become conflated with outdoor leisure – see the later chapter) are
almost completely white. Or white people dominate sports that are associated with public schools and elite universities, such as rugby union, or
sports associated with residual white working-class communities such as
rugby league. White people want to take part in sports that other white
people play, either working-class white folk (if they come from the working classes) or aspirational, bourgeois white folk. They want to demonstrate their distinction and cultural capital, but they are also replicating
the instrumental hegemony of Western whiteness: each choice they
make to take up cycling rather than cricket (a traditionally elite, white
British sport that is now associated with the British Asian community)
perpetuates the structural order and the instrumentalization of sport in
the production of instrumental, hegemonic whiteness.

When non-white people start to become involved in certain sports,
there is a clear white flight from those sports. Basketball and athletics serve as obvious examples of this process. Basketball was created in
the United States in 1891 and heavily promoted to schools and colleges by various Christian associations (including the YMCA, in which
the founder of the game had first come up with the rules) as a healthy
and safe alternative to rougher sports such as football. The sport became
something associated with white, all-American small town life, a part of
the rites of passage of American boys becoming men. Girls danced on
the side, the best boys played, and the boys who were not strong enough
to play for their school were side-lined in the pecking order of school
corridor politics. American colleges had their own basketball teams and
they offered a route for successful white boys to get college degrees while
getting the fame and glory of Varsity. Elsewhere in the world, basketball

was adopted as an amateur, white man's sport in European countries and former European colonies such as Argentina. Professional basketball leagues appeared in the United States and soon white basketballers were making money as sports stars – but the professional leagues offered the chance for poor men from different ethnic groups to play. The increase in the number of African American professional basketballers led to coaches, managers and physical education teachers believing in false racial science and seeking out more African Americans, preferring them over their white rivals because they thought (incorrectly) black people were biologically fit to be better than white people at the sport (Spracklen, 2008). White folk then started to think basketball was dominated by black people, and started to encourage their children to do other sports. African Americans saw a chance for their own children and encouraged them to play basketball. The effect was a rapid change in the perceived and actual racial formation of basketball at elite level – and a drop-off in participation by white people. The National Basketball Association (NBA) does offer a route to success and fortune for a small number of African American men, but the price is the exoticization, brutalization and appropriation of their blackness by the white men who dominate the sports industry, as well as the stereotyping of black masculinity in the white gaze of basketball fans and the shattered dreams of the large proportion of rejected athletes (Hall, 2001).

Athletics is another sport in which the demographic profile of participation has altered in United States and Great Britain. Traditionally an amateur sport with strong associations with elite schools and universities, the success of a few African American elite athletes in some athletics events led people in athletics to follow the same false racialized thinking mentioned in the discussion about basketball. Coaches and educators assumed the success of some black athletes was because of the biological make-up of the black 'race', which they falsely believed to be suited to sprint events (Spracklen, 2008). This led to more black participants in athletics being pushed to the sprint events, while white participants were encouraged to rake part in events where black physicality was not at stake. In turn, the wider group of white people started to believe that sprint events were dominated by black people because of their biology, and not the racism of the choices and structures at play. White people encouraged their children to run in the middle-distance events not the sprint events; on turn, black parents encouraged their children to take part in sprint events, or rather coaches and teachers assumed their black pupils would be 'suited' to those events. Athletics today is still dominated by white people in the West participating as amateurs, but the racialized ideology of biological fitness means the elite side of the sport has developed coaching programmes targeting and nurturing black athletes. When such athletes win medals it is a success for the struggle for civil rights, but it is also a reminder of the ways in which white people have made black people the subjects of their bodies and a mythology of sub-humanism (Carrington, 1998).

The racialization of sports participation and the racist myth of racial biology appears again in the literature on racial stacking. Team sports such as volleyball, rugby, soccer, American football and cricket all demand different physical and mental skills from individuals in different positions in the teams – in American football, for example, the strategic planning and thinking is done by the quarterback, while offensive runners need to be quick and defensive players need to be strong so they can tackle and block. In the 1960s onwards, as more non-white sports participants played team sports to an elite level, they were selected in positions where speed and/or strength was paramount over intelligence – some of these non-white players had started out in thinking positions as junior players but had been moved to the marginal positions as they progressed through the coaching systems (Long, Carrington and Spracklen, 1997). It became clear that coaches were selecting non-white players in positions that reflected a racist understanding of their physicality: black people were being stereotyped as quick and strong, but stupid. Despite the issue of stacking become evident in research in the 1970s through the 1990s, it has continued to appear in professional team sports, especially where racial prejudice and white hegemony remains part of the public sphere (Tatz, 2009).

Sports remain leisure activities in the West where participation is strongly racialized and Habermasian instrumentality is at work normalizing white power. Whether it is the bourgeois classes seeking out aspirational sports that shape their cultural capital, or right-wing, nationalist types finding a space away from the multicultural reality of today, sports participation retains an exclusionary whiteness. Sports participation is predicated on access to social networks, confidence of being accepted, education, peer support and parental sacrifice of time and money. While golf clubs can no longer have signs on the door to keep out black people and Jews, private sports clubs remain exclusionary in their constitutions, cultures and practices – the existing members of private sports clubs can still make decisions to keep their clubs mainly white, mainly male, mainly elite, without formally following written rules that discriminate. Gaining access to private sports clubs, which offer the best facilities and often the best coaching, limits the opportunities of marginalized groups to participate in sport. Decisions on funding programmes in the public and voluntary sector might be guided by more liberal political imperatives, but such funding is limited and often restricted to keeping the existing sports participation rates across the population from dropping any quicker than they are doing. As such, funding is often target at sports and sports clubs that already have participants and access to public funds – this means the cycle of funding rounds perpetuates the previous levels of inequality and inequity in the allocation of money, and the existing whiteness of sports participation in Western countries is reproduced with a new cycle of grants. In England, for example, Sport England historically gave most of its funding to sports such as soccer, rugby union, rugby league, swimming, cricket and tennis – apart from the soccer and possibly cricket, the others are sports that have a long history of being part of the prevailing white hegemony in English society through the low numbers of non-white participants; and all these sports have a history of racism and prejudice and exclusion. The trend in the United Kingdom to focus funding on sports that have had success at the elite level means increased public funding of governing bodies of sports that remain very white. Plans to link funding to successful completion of and compliance around the Equality Standard put in place at the beginning of this century seem more aspirational than operational (Long, Robinson and Spracklen, 2005; Spracklen, Long and Hylton, 2006). With funding slashed for community sports development activity in local authorities, British sports are in danger of reverting back to their traditionally white state after some years of legislative and policy-based equality activism. The same dangerous trends are evident in other Western countries.

Sports fandom

The sports fan is an important social identity in modernity. Being a fan of the New England Patriots or Manchester United or Real Madrid is often, for the individual fans, a strong part of their image and self-expression. Supporting a particular club or a particular professional sport, demonstrating that support through going to watch sports events, or watching on television with other supporters, is an important part of being a sports fan. Being a sports fan gives one a sensed of solidarity, a sense of belonging, in the fact of being with fellow fans or in the sense of sharing traditions, symbols, myths and community (Hugenberg, Haridakis and Earnheardt, 2008). Fans do not need to have played the sport they support, but they do need to have an understanding of the sport's rules, history and cultures. They demonstrate an awareness of which athletes or teams have had the most success, they know some of the great stories of the top competitions, and they recognize the colours and badges of clubs and countries. In many countries some professional sports are watched mainly by men, or have a tradition of mainly male support, so being a fan of soccer in Italy or ice hockey in Canada (say) might provide one with a sense of shared masculinity (Connell, 1995). Historically, many sports were associated with particular classes in white, Western society, and these distinctions might still hold: so soccer in the United Kingdom has this long shadow of working-class identity hanging over it, and many local clubs outside of the Premier League are still dominated by white, working-class male supporters. Where countries are modernizing, being the fan of a particular sport might mark people out as traditionalists or as middle-class Westernized elites: in Japan, for example, during the late twentieth century being a sumo follower or a baseball fan was a political act as much as a personal choice (see also Saeki, 1994).

In research on soccer fans, there is strong evidence that supporting a club through being at the stadium cheering them on is seen as being more proper or more authentic than a fan who watches the team on television (Hugenberg, Haridakis and Earnheardt, 2008). There is related research that suggests there are tensions in modern professional sports between fans that have grown up watching the club because it is their home-town team, and fans who have adopted the club as their club because the club has won a few trophies. In the United Kingdom, this tension has led to fans of Manchester United and Liverpool FC launching campaigns against newer, richer fans of their team, and even setting up clubs of their own for the 'real' fans (Millward, 2011). There are also tensions between fans of existing franchise clubs in sports like American Football and the older fans that supported the franchise when it was based in a different city. Being a sports fan might feel like being a passive consumer, meekly buying the season ticket, the new club colours, the expensive food and drink inside the stadium, but it is also something that fans believe to be an active act of agency on their part. They may have become a sports fan through their father taking them to the local club, or they have become a fan through friends at school or college or work, or they may have come to sports through watching television, but whichever route they took to becoming a sports fan that fandom is their choice (Roberts, 2004) – they will choose to be passive or active fans, they will choose to spend money or stay at home, they will follow their sport's or their team's fortunes with great sorrow and great joy.

These tensions over community and identity are linked to romanticized, nostalgic white imaginary and imagined spaces. Historically, being a sports fan typically involved white people of different classes paying money to support their local sports club. As modern sports were designed as spectacles, people got into the habit if supporting local teams or athletes, or gathering at elite events to show their good taste. For the white working classes of the West, this involved identifying with a club taking part in an elite male team sport, whether it was professional or amateur. The most popular sport was soccer, but other codes of football dominated in certain countries and regions, and not all these codes were overtly professional for much of their history. Other sports such as (ice) hockey, cricket, handball and baseball acted as ways for white, working-class communities to create belonging through local pride, local identity and the exclusion of the Other. For the white bourgeois and elite classes, this popular sports fandom was something they could access and engage in (and many sports clubs had stadia in which poor and rich were allocated separate spaces to watch the sports action). But they also had more exclusionary routes to sports fandom – through following and supporting elite sports such as tennis and golf, and through acquiring the cultural capital to understand and follow such sports. These elite sports allowed elite white people to prove their moral worth – or, rather, prove their insatiable snobbery – by rising above the prejudices of the football crowd to become someone attracted to the aesthetics of individual play. These purposes for sports fandom remain today important reasons for white people's continued interest in sports as spectators. White working-class men (and women) define their right to some local area through their allegiance to a particular club or local sport; in turn, bourgeois white people seek to demonstrate their elite status by gaining white cultural capital from bourgeois sports fandom (still expressed through sports such as tennis and golf); and the ruling white elites use elite white, Western sports such as polo to mark out their elite and exclusionary spaces.

As well as local chauvinism and snobbishness, modern sports allowed individuals and other groups to express crude notions of nationalism through leisure. The urge to compete and be the best in any sport led quite quickly in the history of modern sport to the creation of national teams and international competitions with athletes representing their countries. This sports nationalism was a product of the nationalism of the late nineteenth-century, when European nation-states were using culture and myth to bind their citizens into a common, patriotic cause against the other European nation-states. This was the era of Empire building, of popular belief in the supremacy of white, European, Aryan nations – so it is no surprise that the most energetic nationalizers in sport came from this area, or were in former colonies of Europe such as Australia, New Zealand, Argentina and South Africa. For sports fans, supporting the nation took on huge racist, nationalist significance – it was a means of demonstrating the superiority of the nation against its rivals; it was also a means of identifying with the nation as one's homeland, the land which belonged to the collective of white sports fans cheering on the national team (Bishop and Jaworski, 2003). Team sports such as soccer were the most obvious places where this exclusive nationalism was constructed, as they were popular across all (or most) of the classes of modern nation-states. In the twentieth century, international soccer fixtures offered their white, working class, mainly male fans a chance to escape the uncertainties of their lives and find solace in community and communal identity – if the national side beat the rival side it was an expression of the rightness of the nation and the masculinity of its footballers; if the national side lost it was a disaster brought on the country by players who were not passionate enough when wearing the nation's colours. As soccer spread through the world, it served as a site for nationalism in new, postcolonial nations and other countries where modernity brought globalization (Rowe, 2003). In the white nations of Europe and the white European diaspora, nationalism, racism and national teams in sports became sites of tension between the old white hegemony and the new multiculturalism of the late twentieth-century. Supporting the national side at soccer or cricket or rugby was judged to be a marker of citizenship and belonging by right-wing politicians, even though the sports and the sports grounds remained places for white prejudice and racial discrimination to take place: in the United Kingdom, for example, Conservative MP Norman Tebbitt criticized non-white people in the country for supporting other countries such as Pakistan (the countries from which they had come, or from which their immediate ancestors had come) at cricket test matches. While today there has been much progress in some countries at using national sports teams to promote a diverse, multicultural nationalism (see Bairner, 2001), the suspicion remains that non-white people are not welcome in their own country if they choose not to support that country's sports teams – and in some European countries the fans of national sides in team sports remain exclusively white, and the fan culture remains racist and exclusively nationalist (Sack and Suster, 2000). This white grip on sports fandom can also be seen in countries where white people have lost political power: the South African Springboks remain a national team cheered more by the white people of the nation than their black counterparts, a team where white Afrikaners can pretend for eight minutes that they still control the country (Nauright, 1996).

In the United States, such virulent, racist nationalism has not been so obvious in the history of fandom in its sports scene. This is probably because the nation's biggest spectator sports are ones that do not have well-established international competitions (even if baseball calls its finals the World Series). However, whiteness and parochialism both play key roles in sports fandom in the United States, both in the past and in this century. American fans follow their chosen clubs as a means of demonstrating their Americanness, their connection with the nation-state. American nationalism assumes that the United States of America is a chosen nation, blessed with the favour of God and fighting the good fight against the forces of evil. This creates an inward-looking, parochial nationalism associated with white myths: the importance of baseball and supporting the local teams, the rites of passage of sports in schools and colleges, and the confidence in American power that does not need success in an international sports area to support it. Sports become sites where white Americans fearing their future as a residual, marginalized culture can dream of all-American, white towns and suburbs with white picket-fences and whitewashed walls mark out white America (Bairner, 2001). Sports are sites where rich white Americans can still exert their political and social hegemony, through sponsorship, ownership, management and merchandizing. And white sports fans in the United States can get satisfaction from seeing African Americans bought and sold like commodities in the brutal world of professional team sports, individuals reduced to their physicality, their value as athletes, racial myths made flesh to satisfy the white gaze of the spectators.

Modern sport has become a huge industry guided by legislative and policy frameworks inside nation-states and across the globe. Various countries have legislated against racism in professional sports stadia, and initiative to promote racial equality now exist at a number of national and transnational levels in sport. These initiatives have led to some welcome increase in diversity in sports fandom. But to be a sports fan in most Western countries is to belong to a residual, white male hegemonic culture – sports are where white fans can celebrate their local and national pride, their mythologization of their own place in the nation, and the racialization of black bodies. To conclude this section, it can be seen that the modern, professional sports industry continues to be a place for the construction of white hegemony even where globalization and other socio-economic changes pose a challenge to sport's whiteness.

White people find an imagined community of 'pure' whiteness in supporting a particular club or national sports team (for example the South African Springboks) and this whiteness is only partially challenged by the introduction of black athletes into those teams. White sports fans are (generally, mostly) comfortable with black athletes representing their club or their sport or their country because the globalization and professionalization of sport has turned young black athletes into caricatures of physicality and modern-day slaves. White sports administrators are still in charge of the sports, still defending the white history of these sports even as some black athletes are embraced as exotic 'ringers'. As such, the power dynamics ensure whiteness remains hegemonic and instrumental.

Conclusions

The ideology of modern sports says that anyone can take part in any sport, and the best athletes win through fair competition. Neither of these elements is true. There are two myths associated with these two elements of the ideology. The first myth is that of equal access. Modern sports have never been about promoting equal access: as I have shown, they were designed as elitist practices and developed into practices of elite agency and subjugation of the masses. Sports are not open to all by definition. They are designed to cater for those who can afford to join in, those who have the connections to belong to a club, those who have the right education to be given the chance to pay a certain sport. Sports participation is loaded in favour of the elites who invented modern sports, the white elites of the West, who used sports (and still use sport) to bolster their instrumental, hegemonic power. The second myth is related to the ideology of the best athletes rising to the top. Advocates of modern sport's levelling myth would argue that sports are predicated on a level playing field for everyone. Only the best athletes get to the top. No coach or manager of a professional sport would ever make the decision (on racial grounds) not to pick their best athletes for inclusion in competitions – and also, in the same flawed argument, the existence of stacking, or the dominance of non-white athletes in certain sports, suggests those athletes are 'naturally' the best. I hope this chapter has demonstrated that this second myth is also false. Non-white athletes have been and are routinely not selected, because the racial prejudice of the selectors has often outweighed the commitment to won. And sports fans have allowed themselves to have two incoherent but related worldviews – firstly, that worldview that claims non-white athletes are both unreliable and not committed to their sport/club/country; but secondly, non-white athletes are brutes, suited to the physical nature of certain elite sports arenas.

This chapter has shown how racialization and the construction of whiteness occur in modern sports to provide unequal access, opportunity and outcomes. One of the things that can be noted from this analysis is the intersectionality of the inequalities at work in modern sports: sports are a way of defending heterosexual masculinity and elite and bourgeois classes, as well as whiteness. They are a way of affirming particular gender roles for women and men that stick to the Gender Order of the pre-modern world (Connell, 1995). They are a way of reifying certain body types over others. And they are a way of marginalizing different sexualities or different castes, people with disabilities, people of lower incomes, economic and political migrants, and people from different nationalities within modern nation states. The zombie of the Victorian gentleman, sexless and powerful, able to swing a bat at the crease in the same way as he swings his whip at a native servant in some far-off colony, able to abide by the Bible and by Homer, lives on in sports today.

Policy-makers can intervene and give guidance, and funding can be cut where governing bodies fail to reach levels in various charters and standards designed to promote equality (see Long and Spracklen, 2010), but the real challenge to racism institutionalized in the symbolic boundaries of sports communities will only come about through sports themselves taking a lead on this issue. However, even if the commitment is there to work through the higher levels of these standards, the whole process is predicated on an assumption that sports, as well as tackling racism, are in a position to help social inclusion and community cohesion. In the United Kingdom, although Sporting Equals, the Commission for Racial Equality, the Government and Sport England all believed (at the turn of the century) in the value of sport as a vehicle for social inclusion and community cohesion, this assumption has yet to be proven, despite persuasive arguments and anecdotal evidence from case studies (Long and Hylton, 2002). If the Sporting Equals project was to make a difference in sport, the tension between sport as a vehicle for inclusion and sport as a vehicle for exclusion needed to be fully understood (Long, Robinson and Spracklen, 2005). The Racial Equality Standard attempted to tackle racism and promote racial equality in sport, but by tackling the former one does not necessarily promote the latter (Spracklen, Long and Hylton, 2006). Governing bodies keen on showing their social inclusion credentials by doing development work with people from Asian and black and other minority ethnic communities are struggling to achieve racial equality if they do not change their structures, practices and cultures. This ambiguity about the value of sport persists. In other words, what is at stake is the imaginary community of each sport, the shared sense of belonging that already binds people in every sport to that sport and to other people in that sport. This sense of belonging, the idea that one badminton player shares a sense of familiarity and belonging with another badminton player, is an important way of including people and giving them a sense of social identity. But this process of identity formation is at the expense of outsiders; and in English sport the Asians and black people on the outside, the Other, have been excluded through a process of imagining community in sport that supports and reflects the hegemonic (but contested) status of white, generally middle-class, generally male people. The challenge for those seeking to promote racial equality and social inclusion is to contest and change the symbolic boundaries of the imaginary communities that surround individual sports and 'sport' in England, from the local table tennis club all the way up to the policy-makers in Sport England and the Government. This final point should not be overlooked. The culture of English 'sport' in general, the romanticized mythology of medals and masculinity, needs to be understood as an imaginary community established to exclude the Other. If Sporting Equals and the Standard are to have a real impact, if sports role in promoting social inclusion can be justified, then 'sport', the imaginary community needs to change and reflect the desires and values of what Parekh calls our community of citizens and community of communities (Parekh, 2000) – instead of reflecting the desires and values of a hegemonic instrumentalized whiteness.

This chapter has introduced and discussed the history of modern sport and its importance for the construction and maintenance for whiteness. It has been argued that sports participation and sports fandom are two key sites for this identity-work, where hegemonic whiteness is (re)produced. In the following chapter, I will look at this construction and maintenance of whiteness through the intersection of sport with the media in popular culture.

8
Whiteness and Sports Media

Introduction

It is the summer of the London Olympics and Paralympics. On the walk up to the campus, I meet a feminist colleague coming the other way. I haven't seen since the end of term. 'Been watching the Olympics?' she asks. 'Yeah, right,' I reply with what I think is a sardonic tone (I actually don't have a television and I haven't watched any of it, so it's a true response). 'It's great, isn't it?' she continues. I look at her to quickly see if she actually means what she is saying – she looks like she means it, and there were no tell-tale signs in the intonation. 'Really?' I ask, somewhat perturbed. 'Yes, it's been really exciting watching it all, I've got hooked' she says as she heads past me. At that point I carry on walking, too. About a month later I am out on the hills with one of my closest friends. He's an artist, quite famous in the world of fantasy art and comics – not a natural sports person. When we're out walking or drinking together we talk about all sorts of rubbish, but never sport. And yet, when we're struggling up a boggy path he says to me: 'Did you watch the Olympics? It was incredible!'. 'It was a big waste of money and a way for people to make money,' I reply. 'Yeah, I know, but it was great, though, how it brought people together. And watching athletes from Britain do well. I liked the way we got to see sports that aren't normally on TV. Made a change from football all the bloody time'. 'It was a distraction,' I reply, 'just like the Queen's jubilee. Make people happy, keep them in chains'. 'But it got people being proud in the country, without it being hijacked by the BNP', says my friend. We are out on a day's walk, so I stop being an academic and say something rude about skinheads and Nick Griffin (the leader of the British National Party, the neo-Nazis who have tried – mainly unsuccessfully – to tap into English and British nationalism to win elections). But I'm not happy with what he has said, and clearly I remember it because now I'm writing about it.

121

I apologize to my colleague and friend if I have not remembered the exact words of our conversations. This is not a scientific attempt to gauge the impact of the Olympics as a sports media event on the British nation. All I know is two people who were otherwise sceptical or disinterested in professional sport and sports in the media had suddenly found themselves caught in the emotion and intensity of the coverage of the Olympics and Paralympics. The news-papers had special fold-outs published every day of the events, and pages of coverage in the rest of the sections. The BBC played its role as the voice of the Establishment by reporting reverently and uncritically about every aspect of the events. The events themselves dominated the rest of the television sched-ules. Every news site on the Internet seemed to be dominated by who was winning and who was coming second. Great Britain did well on the medals table and the media reported that the nation had come together to celebrate this, a nation that celebrated its multiculturalism, a nation confident in its modernity. It became acceptable to wrap oneself in a flag, the Union Flag of Great Britain, and be British. Many of Britain's high-profile winners were from minority ethnic backgrounds, seemingly confident in their identity as modern Britons. The two people I spoke to clearly had the 'feel good' factor and had been caught up in tracking the successes and failures, the dreams fulfilled, the passion and the pride in the nation. What I thought I saw in their embrace of the Olympic dream was the power of the media to use sport to make people think they are part of something special, and the power of sport in the media to make people forget the inequities that shape modern life. This summer, unlike 2011, the streets of London did not burn, despite the continued racism of offi-cials and police, and the elitist policies of the ruling class. The Olympics served a convenient need to focus people's attentions away from the real problems of inequality and to hide the white Establishment of Britain behind a veneer of inclusion.

This chapter focuses on whiteness and sports media. This is a com-panion chapter to Chapter 7. It could be argued that this chapter should be subsumed into the previous one, but 'sports media' is a completely different leisure category to sport: most people who watch sports on tele-vision or read about them on the Internet do not watch sports events live, and do not participate. Most people who describe themselves as a sports fan get most of their news and information about sports from the media: they watch sports events on television; they listen to broadcasts and discussions on the radio; they browse sites on the Internet; and read the back pages of newspapers. This type of sports fan might occasion-ally be someone who plays sport or who watches sports events live, but the dominant performative role is of the fan as a consumer. This sports fan follows professional (or high-profile) sports because such sports give them a sense of belonging and exclusion: they belong because they follow the sport and talk about the sport with other sport fans, and they exclude through not liking fans of other clubs or sports, or those who have no interest in following sport. With the advent of the globalization of the sport industry, and the rise of digital technologies and media, it is possible to spend all day consuming sports events, through watching dedicated subscription channels on television and streaming of events on the Internet, and through engagement in social media such as Twitter and Facebook that allow a fan to interact real-time with other fans throughout the globe. The sports fan can feel part of the action in the comfort of the front room, eating snacks and drinking beers, or in the company of others in a sports bar with a big screen television. This is clearly something qualitatively different to participating in sport.

This chapter discusses the way in which sports and entertainment intersect and construct whiteness. I will begin this chapter by demonstrating that sports media are one part of a wider entertainment industry, globalized, commodified and controlled by a small number of transnational corporations, which in turn have a close relationship with the anti-regulation, pro-liberalization trends of modern political parties. Sports play a key role in increasing the profits of such corporations, so sports media have always been exploited by such corporations since the advent of radio in the first half of the twentieth century. This chapter will then proceed to discuss examples from Australasia, North America and Europe of recent and current sports media and the way in which such media over-signifies blackness and makes whiteness ordinary and invisible. I will show that sports media – whether traditional television programming of live sports events, or modern media outlets such as online blogs and discussion forums – are places where instrumental, hegemonic whiteness is constructed, with little room for any communicative resistance from subaltern groups.

Sports media: Emergence and development as entertainment

As discussed in the previous chapter, modern sports transformed from elite leisure activities into mass entertainment almost as soon as they spread out from the elite classes. Professional, modern sports established themselves as the successors of the older spectator pastimes such as racing, bare-knuckle boxing and pedestrianism. Professional clubs in team sports such as soccer built fences around their playing fields, made spectators pay to watch and built stands in which spectators could shelter from the weather. Watching sport – paying to watch it, supporting a club – became a fashion among all classes in the West, and a marker of white, modern masculinity (Vamplew, 2004). The growth of professional sports was phenomenal at the end of the nineteenth and into the twentieth centuries, both in terms of the number of professional sports clubs and leagues, and the numbers of people who wanted to declare their loyalty to a particular club or sport. At first, the passion aroused by sports fandom was fuelled by the rise in sports coverage in the newspapers: first match reports, then daily news and speculation (Collins and Vamplew, 2002; Vamplew, 2004). Alongside dedicated sports journalism in local newspapers – stories associated with the fortunes of the local club or the particular professional sport favoured in the area – there soon arose a practice of having professional (and high-profile amateur) sports discussed and analysed in daily sections in the national newspapers. The newspapers competed with each other for the money of the sports fan, each trying to find angles and exclusive content that would make their newspaper more attractive than the others. Modern technologies such as the telegraph and the telephone allowed journalists to report on matches and events taking place throughout entire countries and beyond. As well as newspaper coverage, sports were the subject of specialist magazines aimed at adults and children: the former allowing fans to discuss the action with their friends and engage in bets; the latter normalizing the novelty of modern sports in the minds of the young. All these forms of print media constructed a public sphere in which sports and sporting success defined local and national belonging, creating a myth of traditions around the novelty of modern sport, making sport a symbol for civilized, white, male Western culture and power.

Sports were big business in the West by the first two decades of the twentieth century: professional sports leagues were established in Europe, North America and Australia; and the print media in every Western or Westernized country promulgated the notion that following sport was fashionable, masculine and modern (Briggs and Burke, 2009). With the commercialization – and commodification – of radio technology came competition between broadcasters for listeners and advertising revenue. Radio technology allowed companies to broadcast live and recorded material to anybody who bought a radio set and tuned in to the company's programmes. Sports events were a profitable part of any station's programming, and soon companies were negotiating with professional leagues, clubs and governing bodies of sports for contracts and permissions to cover their sports events. News reporting of scores and analysis of the day's sports action was complemented with live commentary on events, which proved profitable and popular. Radio defined a nation or a region, or both, and its novelty made it a mass-market leisure phenomenon in the 1930s: people listened to the radio at work and at home, discussing the topics broadcast with friends and family. In the United Kingdom, the British Broadcasting Corporation normalized an elite, white version of Englishness through its use of Received Pronunciation, the national anthem and its formal subservience to the Monarchy, the Church of England, the Marylebone Cricket Club, and the British Empire. In the United States, radio established baseball as the national game, allowing live reports and news programmes on the national scene to be syndicated across hundreds of local radio stations, ensuring national identities were linked to the white, small-town, all-American men who played the sport (Silvia, 2007). In Australia, radio created rivalries between cities where different football codes were dominant, but used those different footballs to create a common Australia of working-class white men battling the system, the British and the Aborigines (Moore, 2000). Radio allowed sports fans to follow their national teams into world cups, test matches, and other international competitions such as the Olympics. The commentators and journalists would report about the success of 'our boys' against the foreigners and would play up national and racial stereotypes – so, for example, cricket commentary would highlight the 'natural exuberance' of West Indian sides (Haynes, 2009).

Newspapers and radio still play an important role today in the formation of the sports-media industrial complex, but the biggest influence on the creation of the complex as we know it is television. As I say elsewhere in this book, since the second half of the twentieth century, televisions have become a part of every modern home (and almost every bar and café), in the West and in the rest of the world where conditions allow their use, a device that has become so natural and ordinary that not to have one is to be deemed abnormal or weird, or otherwise contrary. Sports programming has played a crucial role in the spread and normalization of television. Media corporations pay billions of dollars for the right to broadcast modern, professional sports events. When television first became part of popular culture, the bidding wars for exclusive rights to broadcast sport were either inconsequential or non-existent, as the number of channels was restricted. The development of cable, satellite and digital broadcasting technology has multiplied the number of corporations fighting for subscriptions and shares of audiences and advertising. As with radio, digital television has created a 'gold rush' for popular sports such as soccer, and events such as the Olympics, with broadcasters scrambling to outbid each other to get the lucrative deals. The huge sums involved mean that sports governing bodies are willing to compromise their rules and regulations to make the sports more suitable for television, either through changing the timing of competitions or making the sports more 'viewer friendly' – so rugby union, following its professionalization in the 1990s, has scrapped some of its more arcane rules to make it easy to follow on television; and rugby league in Australia has considered allowing more stoppages for ad breaks.

National and international governing bodies of sport, such as the National Football League (NFL), the IOC and FIFA, still control on paper the governance and direction of their sports: they still elect members of committees, debate rule changes at their councils and engage in competitions to select successful bidders for big events. But the power within all professional sports organizations has become centralized – *de juro* and *de facto* – in the hands of powerful paid executive officers or chairs (Booth, 2011). Important decisions might still be rubber-stamped by the democratically elected delegates at a national or international council, but agendas and recommendations come from elite power-brokers who receive regular briefings from corporate lobbyists. It is a matter of record that some international governing bodies of sport – the IOC and FIFA – have been the centre of bribery allegations and scandals over under-the-radar lobbying and persuasion (Jennings, 2011). But although the influence of sponsors is less obviously illegal and corrupt as the bribes paid to smooth the success of bids for big events, it is still an important source of power in modern sport. Commercialized, commodified, professional sport needs sponsorship to ensure the profitability of the leagues, the sports and the events. Some individual executive officers might benefit illegally from kick-backs, but the public face of modern sports is the power that sponsors (broadcasters and others) gain from putting their money into the system. In the London Olympics sponsors such as Visa and McDonald's found themselves able to dictate exclusive rights to usage and services in the Olympic Park. Sponsors gained the best seats for their clients and staff, and had access to the fast lanes reserved for the use of IOC throughout the capital. For all the supposed multicultural diversity of the London Olympics (the spectacle of the Opening ceremony with its 'new' Britain, and the feel-good factor reported in the press), the fact is that the actual local, working-class residents of East London, a truly multicultural demographic, found themselves unable to access the park because of the costs of the events, and unable to drive through their streets because of the corporate sponsors and IOC bureaucrats cruising through the reserved lanes. The only regeneration of the area has come in the form of the Westfield Centre, a huge shopping mall privately owned by a transnational corporation, filled with retails outlets owned by big corporations.

With globalized cycles of bidding and hosting, the impacts and continuing legacies of sporting mega-events like the Olympics are of obvious interest to academics and policy-makers. The Olympic Games have become a symbol of the globalization of sport, replacing the amateurs of Baron de Coubertin with the globalized, professional travels of today. Countries compete for medals, but the main competitions are those between the mega-star celebrity athletes, trained at high-altitude camps, supported by millions of dollars of coaching and science. The Olympics have become the subject of billions of advertising and marketing deals, construction projects, television rights and political lobbying – the global audience for the Games attracts the attention of huge corporations looking to increase market share or break into emerging territories. To accept the Olympic Games into a city is to demonstrate that city's global importance, its modern society within a global community of modern cities. Four host countries, the same urge to be seen to be part of the globalized society of modernity is evident: for Spain, Barcelona 1992 followed the emergence of democracy after the death of Franco; for China, Beijing 2008 symbolized the arrival of China as a geo-political world power. As well as supposedly beneficial policy connections between these mega-events and tourism and leisure, such as increased profits in host-city hotels and legacies of facilities available to promote participation in sport, there are local and global policy consequences on tourism and leisure that challenge the notion of benefit (Hiller and Wanner, 2011).

Modern sports are now effectively controlled by a small number of transnational corporations, who have been allowed by national and international political institutions to develop a neo-liberal global capitalism that benefits their shareholders beyond anything and anyone else (Horne, 2006). The sports-media complex is part of a wider entertainment industry that has commodified global, Western popular culture. And as I say elsewhere in this book when I discuss and analyse entertainment and popular culture, this means that the sports-media complex promotes the interests of a narrow section of elite, white Westerners. The particular case of modern sports can easily be seen in the light of this general case. Modern sport is a product of white, Western modernity, something that has been developed to sell an instrumental myth of individual skill while maintaining the power of white hegemony. Sport is a way of giving people without power, the working class and the marginalized, something to keep them occupied and distracted. If people are too busy gambling on sports and cheering sports sides on and talking about sports and dreaming of sports they do not dream about challenging the status quo and their lack of access to power. Ever since the Roman Empire, elites have used games and sports to keep the oppressed masses from rioting – and this trick is well known to any elite, white, Western schoolboy who has sat through Classics at his private school (see Spracklen, 2011).

The sports-media complex is maintained by a series of professional sports leagues and governing bodies that were designed by white bourgeoisie and white elite classes. These organizations have changed through the process of globalization, but they remain sites of cultural resistance to diversity, multiculturalism and equality. As Western societies have become more democratic, diverse and equal through laws and policies, sports organizations have remained places where older white men (on blazers) have influence. Where international organizations such as the IOC and FIFA have had to give power to representatives of countries outside the West, this has been done slowly and reluctantly, and has only succeeded in allowing Westernized elites from such countries to gain access to the corridors of power (Booth, 2011) – and ultimate executive power has still resided in a white, Western elite of officers. These sports bureaucracy elites are one half of the sports-media complex. The other half is the entertainment industry itself, which is dominated by Western interests and the shareholders and CEOs of white, Western corporations. Like the white people in control of the governing bodies, the white people who run the corporations that sponsor modern sports work are central to the protection of white, Western power: in this case, the power to use and preserve the historical advantages of America, Europe and the West against the rest of the world. Diversity and equality might be used by the sports-media complex to try to make money, or to try to meet equality guidelines and laws in countries that have good measures in place, but the purpose of the sports-media complex remains one that normalizes white privilege and the historical inequalities that remain unchallenged.

Examples of whiteness in the sports media

Rugby and big polynesians

In Australia and New Zealand (and to an extent in North America) the two versions of rugby have become sports where non-white players of Polynesian and other Pacific Island descent have been successful. This success if mediatized through the reporting and analysis of professional rugby leagues, professional clubs and big games such as international tests and the National Rugby League's State of Origin series. Huge Polynesian rugby players, often with tribal tattoos, are presented to television viewers as dangerous, tough men with the 'natural' build to be the best rugby players (Hope, 2002; Scherer, Falcous and Jackson, 2008; Scherer and Sam, 2012). There is an assumption in rugby union and rugby league that Polynesians are suited to rugby because of their 'racial' distinctiveness, and in New Zealand, this has led to white parents discouraging their children from taking part in rugby alongside Polynesians, who are stereotypically defined as being naturally bigger and stronger than white boys (as it is mainly boys) of the same age (Grainger, Falcous and Jackson, 2012). In my time spent researching discussion of rugby league online among fans, I have noticed that discussions about the growing dominance of Polynesian players in Australian and New Zealand rugby league are always associated with their status as biological Others: the normalized, racialized physicality of Polynesians accounts for the growing numbers of Polynesians and other Islanders in the game in Australia, at junior level and at senior level. This is of course based on a myth of racial difference: a myth that Polynesians are genetically stronger and genetically different to the white race (a myth that appears in John Entine's infamous book *Taboo* – see Entine, 2000). The rise in the number of Polynesian professional rugby players in league and union is a fact, but one that has more to do with the way both rugbies have developed in Australasia, and the way in which professional sport has become a way for poor non-white New Zealanders to become successful.

What is notable in Australian rugby league is the trend for young New Zealanders of Maori and Island origin to leave New Zealand for a chance to play rugby league at an Australian professional club or school with a strong record of rugby league success. There is one professional rugby league club in the National Rugby League competition based in New Zealand, but the rest are all based in Australia. Rugby league in New Zealand was historically the version of rugby that was more open to non-white involvement than rugby union. The latter was the game of the white European settlers, the game of the white elites and white middle classes. Rugby union became the national game of white New Zealand. The game was infamous for its exclusive policies and practices, both formal and informal. The New Zealand rugby union authorities sent teams to South Africa at the peak of the protests and boycotts over the racist Apartheid regime. The game entrenched itself in the private schools and universities of white New Zealand. When non-white people did want to play rugby union, the game racialized its structures through promoting Maori teams based on blood and heritage, which served for many years as a way of preserving the mainstream of New Zealand rugby union as white, and ensuring white power remained privileged.

It is no surprise then that historically rugby league became associated with poor white and Maori/Islander participants – partly because it was the only form of rugby they were encouraged to play, and partly because the hegemonic white culture of New Zealand was invested in rugby union. The result of that segregation has been the dominance of Maoris and Islanders in junior and local rugby league clubs, and the continued strength of rugby league in poor, Maori/Islander communities. So it is inevitable that the dominance in junior clubs in New Zealand is repeated in the players who go to Australia to seek their fortune – amateur rugby league players cross over to join rugby union clubs as well as rugby league clubs, especially since the professionalization of rugby union removed some of its institutional and cultural racism. White coaches and administrators in professional rugby league and rugby union, who hold to the racist notion that these players are naturally strong, sign-up these young New Zealanders by the dozen (Grainger, Falcous and Jackson, 2012) – they give them chances over other new signings, and the media and fans turn the new players into exotic, Othered celebrities, taller than normal men or like Superman, faster than a speeding bullet. What is now happening is that the sports-media complex has generated a myth about the dominance of these non-white players. They are not dominant because of racism and poverty, or because of the racism of scouts and selectors and coaches, but because they are 'not like us'. The sports-media complex of commentators and fans has turned these players into hyper-masculine beasts, foreign gladiators given an opportunity to entertain their white masters before their star fades and a new generation of young men replace them. In rugby union in Australia, the whiteness being privileged and preserved is a residual culture of white Britishness; whereas rugby league in the country is the game of the white working-classes, which has become the main game of the sports-media complex in New South Wales and Queensland. In New Zealand, both rugbies have become more like each other in many ways, but union is still associated with white settler nationalism and the elites and league is still portrayed as a sport for 'poor', 'rough' men from marginalized communities. In the intersection between the sports-media complex, nationalism and the politics of 'race', big Polynesians have become a fetish of white instrumentality.

South park and basketball

The makers of the American animated comedy show *South Park* have deliberately set out to shock and offend viewers and the wider public sphere. The programme is set in the small town of South Park and shows the reality of modern American life and American politics through the eyes of its four main characters: the schoolboys Stan, Kyle, Eric and Kenny. The show started out in its pilot episode making fun of the American obsession with conspiracy theories and UFOs, combing pointed satire with crude farce and slapstick. It quickly evolved into a show that has provoked reactions from left-wing and right-wing critics alike (Gournelos, 2009). The show finds dark humour in offensive behaviour and situations and responds to the news of the week (the show is created in season on a week-by-week basis). At times, the shows have cleverly critiqued modern America's rampant commercialism and individualism, typically whenever the egregious Eric Cartman is the focus of a story. The show's writers have attacked corrupt politicians and the notion of the American family, and have mocked Christianity and cults such as Scientology. In one episode the show's writers foregrounded the irrational and arbitrary nature of racism and prejudice by having Eric Cartman become the leader of a neo-Nazi campaign against ginger-haired people ('Ginger Kids', Season Nine). Other anti-racism themes have appeared sporadically over the years as the writers try to attack the right as much as the left – but *South Park* has a reputation for being fiercely libertarian and taboo-breaking that is based on its continuing use of racial stereotypes as much as its politics of individualism (Hughey and Muradi, 2009). The episode 'Mister Garrison's Brand-New Vagina' from Season Nine is a spoof of the widespread use of cosmetic surgery and the notion that such practices will make people feel better human beings. In the episode, we see the white Jewish schoolboy Kyle trying out for the state basketball team, and struggling to make any impact among a number of stronger, taller, black opponents. The coach who doesn't select him tells him Jews can't play basketball – a racial stereotype that appears in American popular culture and in other American comedies (it has been used about the character Howard in *The Big Bang Theory*). Of course Jews can play basketball – they do play basketball, and have played basketball, and in the 1930s, when American Jewish men played sport as a way of seeking to escape poverty via one of the routes open to them, visible Jewish involvement in basketball was part of a stereotype that suggested Jews were 'naturally' suited to basketball (Jaher, 2001). Now it might be that the writers know this and are making fun of the white Americans watching the show who think like the coach, but if the joke is a double-bluff the subtlety is lost on me. Kyle accepts the truth – he is white, and Jewish, and therefore he can never play for the State alongside the black students. But there is a way for him to play: he gets cosmetic surgery and becomes black, with a bigger biological structure and black skin. This allows him to become a successful basketball player, for a few days, but when he is playing the surgery fails and his legs break. By the end of the episode, Kyle is back to 'normal' as a white Jew, and learns a lesson that he should not try to be something he isn't: basketball is not for white boys who have businesses to run and countries to control. Everybody has their place in life and that is the normal way of things: white people might not be good at basketball but they hold every other card, have every other advantage in life, through the history of exploitation and their continued control of the country's resources (and the resources of the West and much of the rest of the world); black people, well, they have their brute strength. *South Park* seems to reinforce what most white Americans believe about blackness and sport by failing to show the myth as anything other than real.

The episode's story is a stupid conceit, but one that reinforces the notion that African American, black people, just dominate basketball because they just are more physical than the white Americans. Basketball fans in the seats watching the big matches might actually be more sophisticated in their critical thinking, and might dismiss the biology of 'race' – but in the wider sports-media complex basketball serves its role as a marker of whiteness and blackness. Basketball is not the only sport in the American sports-media complex where this retreat by white people is normalized as biology, but it is probably the most studied one (Hall, 2001) – and it is one where the myths continue to exert a phenomenal power, making white people feel glad they are not too brutish, making white people see black athletes as commodities, and making black people see a limited number of opportunities to create a sense of their own self-worth. The myth of black physicality gives white America something to say about the history of exclusion and slavery: 'we might have exploited these people for hundreds of years, but look, aren't they good – and happy – shooting hoops?' This racial politics puts everybody in their place and in a hierarchy of power: white Americans are at the top because they are clever because of their racial genetics; black people are at the bottom because they are stupid, but they have the consolation of their brute strength. The myth that Jews don't play basketball isn't merely a clever trick to protect white instrumentality – it also taps into the ant-Semitic notion that Jews are not proper (white) Americans, not properly part of the (white) American nation (an additional irony given one of the writer's Jewishness). The show's writers argue that they mock everything, that they have a right to have freedom of speech and actually they are cleverly attacking both sides of the culture wars in America, and actually they are not libertarian Republicans but have a more nuanced politics (Gournelos, 2009). But on basketball they are resolutely with the (white) American mainstream, nodding and winking at black people and Jews.

Soccer, Eastern Europe and England

Soccer, association football or just football to its adherents, is undoubtedly the national sport of England: the men's game is a fully professional sport with big clubs, highly paid players and millions of fans who consume the information and images produced inside the sports-media complex. As discussed elsewhere, soccer as it is known today was invented in England, and its early globalization was due to the sport's popularity as a mass participatory sport and as a spectator sport. Up to the 1970s and into the 1980s, English soccers players and fans were predominantly white. A few black footballers had become professionals since the end of the previous century, but until the 1970s such black players were sufficiently rare to be feted as exotic outsiders. From the 1970s onwards black players started to break into the professional ranks. At first black players were stacked away from the crucial decision-making positions (in midfield and in goal), getting picked on the wing or as centre-forwards. Most of the black footballers in the 1970s and 1980s were black British men, youngsters born and bred in working-class districts, as English as their white counterparts. However, these players faced an enormous amount of racism from fans. As black players started to enter the professional game, professional clubs remained bastions of working-class, white Englishness: their fans routinely used racial taunts and racial chanting to intimidate black footballers. In the media, the racial taunts were at first ignored altogether, then, when black players started to complain, the sports media questioned the seriousness of the problem (King, 2004). Former players (white) proving expert commentary and analysis made light of the racism stemming from the terraces, as did the owners of clubs when interviewed: and black players faced enormous pressure from their team-mates to comply with the standard line that it was an isolated problem caused by a few 'idiots' (Long and Spracklen, 2010). Commentators often made black players exotic outsiders by racializing their backgrounds, even when those players were selected to play for England (King, 2004). This happened in other professional sports in England in the same historical period – so black rugby league players and black athletes were routinely identified as smiling stereotypes, making the work of performance look easy and natural, and they were the subjects of discussions about their natural talent in ways that white athletes were never discussed (Long, Carrington and Spracklen, 1997; Spracklen, 2001). The mediatized discussion of racism as a non-problem became the dominant discourse in newspapers, radio and television. But the racial abuse continued: bananas thrown on the pitch, for example, along with monkey chants (King, 2004).

Eventually, action by fans, racial equality activists and politicians led to a number of laws and policies that set out to get rid of racism on the terraces (Long and Spracklen, 2010). These actions were related to the wider issue of hooliganism, and their relative success was partly responsible for making soccer an attractive investment to News Corporation when it bought the TV rights in the last decade of the twentieth century. News Corp worked with the football authorities to promote soccer as a respectable, more equal and diverse sport suitable for all the family – but they still marketed soccer as the white, working-class English game through hyping the sport in News Corp owned newspapers such as *The Sun* (Vincent, Kian, Pedersen, Kuntz and Hill, 2010). The sport itself became more diverse on the pitch at the turn of the century, especially as soccer clubs pursued the talented black British players in working-class districts of England (and talented black players from abroad. The England team has had a long line of black players playing for the country. But the fan-base, while not as white as it was, still does not reflect the diversity of the country. The sport is sold in the media as being truly multicultural but the whiteness of its fan-base remains telling, especially away from the big cities, where minority ethnic populations are absent from the fan-base of local clubs. Soccer is still the place where white, male privilege is performed, especially now the old working-class supporter (marginalized and residual) has been replaced by the corporate fan (Millward, 2011). And other mediations of soccer reflected the white nature of the English game: for many years, the celebrity broadcasters leading the presentations of the sport on television have been white (nearly all of them ex-professionals), with black presenters being a minority. Racialized discourses remain popular in discussions of foreign players in the media – the natural talent of Africans being a particularly egregious example, though the media normalizes race through discussions of black British players' skills and the leadership qualities of white British players (Carrington, 2010a).

In the campaign against racism in soccer, a number of organizations were established to work with football clubs and the governing bodies to get them to take action against racism. This work has led to anti-racism campaigns being promoted and publicly backed by the soccer authorities and by particular clubs. At big games on television, adverts are seen around the pitch, or players run on wearing training shirts with anti-racism messages and logos. In recent years, people in the sports-media complex have declared the diversity of the professional players in England as proof that racism has been defeated – and journalists following England into Eastern Europe have highlighted the overt racism that continues to be shouted at black English footballers by fans in places such as Poland, Russia and Serbia (Kassimeris, 2008). In reporting the racism in Eastern Europe, journalists make the comparison that England is multicultural, less white than the poor white working-class countries of the old Communist East. Racism occurs but it is far away, caused by rough people who live on the margins of civilization. This is the racism of nations perceived by the hegemonically white English as being less white, nations that show an inferior kind of whiteness, an identity that is subordinate to the instrumentally dominant whiteness of the West. In playing in international competitions with these countries, the England soccer team gives the sport-media complex many opportunities to prove the superiority of England and the barbarity of Slavs: just as the matches with Germany give the English media the chance to play on the wars of the last century, national matches against Eastern Europeans allow the press to dig deep into the mythical archetypes of the criminal and illiterate Ruritanias on the edge of the map. These are people who the English white middle-classes fear in their everyday lives, the people who lose their white status when they arrive in England seeking work and a better life. Demonstrating these people are beasts in their home-land through reporting on racist abuse at football matches serves the purpose of the instrumental logic of bourgeois prejudice.

There is no doubt that the racism evidenced at some England soc-cer matches in Eastern Europe is shocking, and no doubt that poor white, male soccer fans in countries that are struggling with a host of political and social problems use their sports fandom to publicly pro-mote racist, fascist and nationalist ideologies (Kassimeris, 2008). But the bigger shock is the failure of the international governing bodies of football to take decisive and firm action to ban countries from inter-national competitions if such racism occurs. Football's authorities care more about preserving the value of their sport in the media – and as long as the broadcasters and sponsors are happy they take no action.

For English soccer journalists, this is proof of European perfidy against the English. But the case of John Terry I mentioned in the introduction to this book is proof of English soccer's failure to tackle racism. Despite the campaigns aimed at fans, there is racist abuse on the pitch, and institutional racism in soccer's governing bodies and clubs: black players struggle to get jobs in the game, and the bureaucracies of soccer are institutionally racist. And black players are still castigated for daring to speak out against the mediatized image of the game as equal and fair. When two black players refused to wear an anti-racism tee-shirt they said they did not believe soccer's authorities treated allegations of racism seriously (a position that rings true given the poor outcome of the Terry case), it was their decision to refuse to play along with the mediated version of soccer as racism-free: but their actions were criticized publicly by their managers and others in the media as failures to support the campaign and do as they were told (http://www.bbc.co.uk/sport/0/football/20032839, accessed 30 October 2012). White men still control English football and its mediatized representation, and they still tell black men to keep their mouths shut about the ingrained racism in the sport.

Conclusions

In the first section of this chapter, I explored the historical development of the sport-media complex, which is part of the professional entertainment industry. One of the enduring myths of this industry is that individuals who have talent naturally rise to the top in some survival of the fittest. The Darwinian metaphor is an easy one to make, but the logic is flawed. In the theory of evolution through natural selection, the natural (genetic) adaptations that are selected are blind chances in the chaos of the physical world: that is, the conditions in which one adaptation might prove more advantageous in terms of fitness and survival are not controlled by agents in the struggle for survival, nor are they predictable in any way. What might be the best genetic combination for the survival of a species in a time of drought might well be something that ensures a species is made extinct in a time of ice-sheets covering the land. Survival of the fittest in an evolutionary sense is a random and directionless happenstance that produces the epiphenomenon of progress and direction. In the modern world, in the entertainment industry, who becomes the professional actor or the professional baseball player is entirely dependent on constraining factors and controls put in place by the hegemonic elites who control late modernity. There are millions of potential athletes and millions of potential actors, and they live in all parts of the world and in all manner of social classes and groups – the bodies and minds needed to become a success in sport or on the stage are not rare. Professional sports that demand an enormous amount of practice or equipment are only available as careers for those who have the money and access to the right facilities from an early age, as I have said in the previous chapter, so these sports preserve and perpetuate hegemonic whiteness when fans consume the sports action. Where black players emerge onto the professional sports arena, their blackness is noticed by the media and the sports fans, who then assume those black players are naturally talented. This then leads white consumers of sports media to think the sports that have black players are sports that are played by black people, and lead black consumers of sports media to see these sports as a way of getting their sons and daughters some success in a world where many opportunities for growth and development are limited for minority ethnic populations in the Western world. The previous section of this chapter demonstrates how the sports-media complex over-signifies blackness and makes whiteness ordinary and invisible, whether it is rugby in Australia and New Zealand, basketball in the United States or soccer in England and Europe.

Sports media – whether traditional television programming of live sports events, or modern media outlets such as online blogs and discussion forums – are places where instrumental, hegemonic whiteness if constructed, with little room for any communicative resistance from subaltern groups. The London Olympic and Paralympic Games of 2012 impressed themselves as sports mega-events on the minds of millions of television watches, bloggers and newspaper readers. In the United Kingdom, the blanket coverage allowed politicians such as David Cameron and Boris Johnson to proclaim the country was confident with its multiculturalism and with its nationalism, that it was possible to accept new narratives of belonging as long as they were wrapped in the flag that had adorned the streets earlier in the summer when the Queen's continued existence was 'celebrated'. This is the politics of divide and rule, setting different minority ethnic communities against each other by measuring their loyalty to the white British Establishment and their closeness to the practices of instrumental whiteness – so black athletes become British through embracing the flag and winning their medals, but poor Muslims in the north of England are the subject of vilification and racism in the media because they have no desire to be 'loyal' to a political elite that engages in post-colonial warfare in Iraq and Afghanistan. In the avalanche of instrumentality surrounding the mediatization of the Olympics, gold medals became symbols for imperial success and loyalty to the Crown, to the Army and to the coterie of public-school educated white men who dominate the Government and the finance industry. With every celebration of sports success, people were told by the media that they could feel good about Great Britain (or England, or the United Kingdom) and the country's place as a world power. Britons took pride in their country's importance as the origin point for modern sports and took pride in the fact 'we' showed 'them', the rest of the world, that we had some highly trained athletes who were better than other athletes – 'we' might be fighting a pointless war in Afghanistan, 'we' might have had our futures taken away from us by global capitalism, 'we' might have missed the chance to get rid of the unelected House of Lords (a key part of our political system and a completely undemocratic part that is bound to the equally undemocratic and feudal role of the Queen) but at least 'we' won all these shiny things.

9
Whiteness and Everyday Leisure

Introduction

It's Monday morning and I am in the shop at my local railway station. It's 7.30 in the morning but for some reason there are no Guardian *newspapers left to buy. I ask for my usual small latte and ask the bloke behind the counter what's happened to* The Guardian. *'Just sold out of them today,' he shrugs. They never have many at the shop. It's a small town, the end or the railway line, and there are never that many liberal intelligentsia on the 7.47 out to Leeds – so the shop only has three or four copies of the paper. I ask him if he could put a copy of the paper under the counter for me in future. He agrees but says 'you do realize the papers are full of rubbish, don't you?' I know that! I know even* The Guardian, *paper of the bourgeois liberal elite, is awfully white and awfully partial, and I despair of its lack of detail on world news, and its obsession with a narrow, metropolitan middle-class Southern elite culture. But I need my daily paper because I'm a crossword addict. Not the quick crossword, the cryptic one. If I don't get to have a go at the cryptic crossword on the way to work I feel I'm missing something, and I get fractious. I do the crossword because it exercises my brain, and I love the beauty of the solutions (know-all rascal met with disaster [5,4]). But I am aware of two things when I do these crosswords: firstly, I am doing the crossword because I am a professional commuter working in a professional job, and maybe I am just replicating the accepted behaviour of the middle-class white English professional on the way to work; and secondly, to solve the clues it is necessary to have a general knowledge of very specific things in white, Western (English) elite culture – the paper's cryptic crossword of 10 August 2012, for example, depended on an extensive knowledge of TS Eliot.*

In the preceding reflection, four informal, everyday leisure activities can be highlighted: having a chat with someone; drinking a coffee;

reading; and doing a crossword game. Everyday leisure activities are those commonplace leisure activities we might not even notice we are doing, or if we are doing them we might not realize they are leisure activities. They are things we do to fill our free time when we are not actively altering our behaviour, but also things we might not think of as leisure (unlike, say, reading a book or watching television, or playing sport) – it is a leisure that is overlooked or taken for granted, things such as eating and drinking, talking and socializing, surfing the net. Everyday leisure activities take place throughout our working lives as well: we are just as likely to go on Facebook at our place of work; we go to lunch; and we get satisfaction from having a chat with our work colleagues about stuff. Some everyday leisure activities are individual acts, or Habermasian communicative actions between two or more people. In stopping to have a chat with my friend at the railway station shop I am choosing to spend some of my free time talking to him – and he can choose to engage in the conversation or not. But the other three everyday leisure activities I have identified are more problematic: I do not need to buy a coffee; I do not need to buy a newspaper; and I can only do the crossword if I buy the paper. Everyday leisure, then, while being informal and grounded on agency, is shaped by the late modern turn in leisure and culture to commodity, commerce and consumption. As I will show in the rest of this chapter, everyday leisure, like any other form of leisure, is a construct subject to the push and pull of the struggle in late modernity between Habermasian rationalities: the communicative desire of individuals using their agency to make social reality (what Habermas calls the lifeworld) and the instrumental power of global capitalism, instrumental whiteness and state networks trying to colonize the lifeworld.

In most theorization of the everyday, the connection is made between individual agency and corporate power: Lefebvre (1991[1947]) is sceptical of everyday leisure when it is a function of capitalism. Debord (1995[1967]) suggests that in such a situation of increasing commodification and centralization of power, there is an imperative and an opportunity to express opposition to such instrumentality through subversion and through spectacle. However, Lefebvre (1991[1947]) also demonstrates the importance of the everyday rituals and routines of individual lives – and the importance of (communicative) leisure and leisure spaces in allowing transformative, counter-hegemonic agency. What this suggests is that everyday leisure can be an activity – a space – where oppositionality might occur. Oppositionality is the way in which individuals, subcultures, counter-cultures and other counter-hegemonic movements reject the restrictions of instrumentality and express their refusal to conform as passive consumers. This might be seen to work against hegemonic whiteness. But as I will demonstrate in this chapter, oppositionality is just as problematic in terms of normalizing whiteness as instrumentality. This chapter will involve both secondary analysis of existing research and some primary research around everyday leisure activities such as shopping, drinking, dancing, reading and interacting on social networks. The chapter will span the Western world, with examples drawn from everyday leisure in North America, South Africa, Australia and Europe. In the following section I will discuss and highlight a number of examples of everyday leisure. I will suggest that everyday leisure activities are particularly problematic precisely because they are viewed as everyday mundane, normal, routine leisure choices. None of these everyday leisure activities are necessarily white: however, the whiteness of these activities is (re)produced in the cultural capital these activities construe. In the section titled 'Ethnic food', I will focus on eating out and the phenomenon of 'ethnic food', drawing on existing research and new data collection to examine the ways in which eating such food construct both a foreign Other and a white Us.

Everyday leisure in the west

Leisure and sports scholars have started to explore the intersectionality of social inequalities in everyday leisure and sports activities. At its most basic, intersectionality is a statement of the obvious: 'race', class, sexuality, disability, gender and faith divisions create social inequalities that cannot be isolated from one another in sociological analysis. That is, all of us have a sense of our own place and the place of others in the social structures around us, and we all understand that those social structures operate collectively on us. Intersectionality is often used vaguely to refer to the way in which traditional social structures and social inequalities combine to create additional inequalities and inequitable relationships. For example, in everyday leisure, men have more freedom and power than women because of their control of money, time and resources. But white women will have more opportunities for leisure than black women, and middle-class women will have more leisure opportunities for leisure than working-class women (Green and Singleton, 2006). In this sense, intersectionality recognizes the complexity of individuals, social structures and wider society – and the cumulative negative impact of social inequality. Intersectionality is also used in a more reflexive way to refer to the many different ways in which the different inequalities overlap – this often has a negative consequence, but not always, some-times the consequences are neutral or even positive. In everyday leisure such as going out on an evening (playing bingo, perhaps, or having a night-out with the 'lasses'), working-class status combined with sub-ordinate femininity might actually mean more leisure freedoms for working-class women than middle-class women (who have to behave in certain restricted ways in their leisure time). In this section, I will explore whiteness and everyday leisure through the lens of intersectionality. There are a number of everyday leisure activities I could explore, but I will limit myself to examining drinking, shopping, reading, dancing and socializing on online networks.

Drinking

In Western countries, where alcohol is permissible by law and tradi-tion, and a part of popular culture, the consumption of alcohol plays a crucial role in the social construction of gender (Lyons and Willott, 2008). Historically, drinking in bars was viewed as acceptable for men and unacceptable for 'decent' women. Women who frequented bars were deemed to be prostitutes or immoral characters of low repute. In certain circumstances women were allowed by social conventions into these male spaces: some bars had outer and inner areas, and women could drink in the outer area if they were with their husbands. Where women were accepted into bars, often they were limited to a small range of acceptable drinks: weaker ales, for example, or mixed cocktails, or no-alcoholic (soft) drinks. For men, bars were spaces where they could be away from their wives, where they could go tell jokes, talk sport, play pool and drink beer. They could prove their masculinity by keeping pace with the other drinkers, and like the women they had a small range of acceptable drinks: beer, spirits, coffee (in some countries that had a tra-dition of coffee drinking), but never any other non-alcoholic drinks. The structures defining acceptable and unacceptable gender roles were social structures, and although they have weakened in some places, drinking alcohol in bars is still viewed by many people in many countries as a male preserve. Women and men learn the correct gender rules and roles through their everyday lives, and quickly learn how to behave according to the social structure of gender in their alcohol use and bar frequenting.

People choose to drink to excess, for example, because they feel it makes them a part of a crowd of like-minded socialites, and the pleasure of the bar and pub is the free, public discourse that takes place there. On the other hand, this form of dark leisure is shaped by the forces of instrumentality. People drink to excess because this is the only outlet for their alienation in a rationalized society. They drink particular brands of alcohol because those are the ones globalized and sufficiently commodified to have massive marketing campaigns. They buy the drink and think they are cool, but they are still paying their money to the corporate machine. People drink at certain times, in certain places and alcohol of limited strength, because of the restrictions associated with legislation and licensing. So drinking to excess is not only a communicative release, a pleasure and a choice in a world with little choice but also an instrumental choice hemmed in by the rules and policies that govern public spaces and the hospitality industry.

So far I have discussed drinking alcohol. But the concept of social drinking or drinking as everyday leisure crosses cultural, historical and special boundaries. In late modernity, drinking a latte has become a marker of whiteness, Westernization and bourgeois sophistication. In developing countries such as South Africa and Kenya, local drinking practices have been swapped by the new elites in those countries for the taste of Americanized coffee. The new black economic elites in South Africa and Kenya have adopted white, Western, middle-class styles – through going to Starbucks they are becoming white in the same way the black political elites have adopted golf clubs in Kenya (Wrong, 2009). In South Africa the old white ruling class has distanced itself from such commodification of coffee drinking and retains an old colonial obsession with private clubs and hotels (Bond, 2000). In the West, going out for a latte, meeting friends or drinking alone (surfing the net at the table, or reading or working), is a commonplace of everyday life. What started as a middle-class marker of distinction has become something of a mass market, but there are distinctions at work, nonetheless. White elites in America and Europe choose to drink their coffee in independent shops or small chains, or other more exclusive venues such as hotels. The white middle-classes prefer independent but settle for a chain that pretends to be independent – Caffe Nero rather than Costa, for example. Or they spurn coffee altogether: one trend emerging in the United Kingdom in 2012 was the idea of 'builder's tea' – an expression used by white elites to refer to milky tea with sugar, used in an awfully patronizing way by rich people who order the tea alongside their full English breakfasts in greasy spoon cafes.

Cafes, bars and pubs are part of the hospitality industry, often owned by transnational corporations such as Starbucks and subject to the instrumentality of those corporations. They want to take our money and make us feel we are making free choices when we enter their doors. It is a con. But these corporations, and the people who own independent bars and cafes, are catering to a communicative leisure desire. Humans enjoy being in the company of other humans. We enjoy taking part in the public sphere, talking about stuff, setting the world to rights, and having a drink with our friends. Whether it is a coffee house in Istanbul or one in San Francisco, or whether we are drinking tea or fermented milk or beer, having a drink with friends is an important part of communicative leisure. We use the public sphere of cafes and bars in the same way the Enlightenment philosophers used the coffee shops of eighteenth-century Paris and London. We feel we have the freedom to think things through, to debate and discuss, in the pleasant and secure company of like-minded people (Jerrome, 1984). Everybody chooses how and where they engage in such activity. It may be that colleagues in a workplace use a tea break to gossip and unwind from the routine, cursing their managers for their ineptness; or it may be that somebody drinks at the local pub to meet others who live in the local community, to engage in banter and to play bar games. The permutations are as diverse as the diversity of humanity. However, drinking also serves as a way of creating exclusion and normalizing hegemonic whiteness. In the United Kingdom, drinking alcohol is viewed as a popular, long-standing British cultural practice: in sports clubs, players and fans drink alcohol after the game; in workplaces, colleagues who work together will bond through drinking alcohol after work at a pub; and drinking alcohol is considered a mark of coolness among young people (Hollands and Chatterton, 2002). However, such normalization of alcohol-drinking serves to exclude religious-minded British Muslims from taking part in leisure activities where they might feel uncomfortable. Black British people and secular British Asians hence become more able to be accepted as British by white friends and colleagues through adopting white British alcohol-drinking habits (Archer, 2001; Jayne, Valentine and Holloway, 2008).

Coca-Cola ('Coke') is a form of sugary, carbonated, flavoured water that has become emblematic of the globalization of American brands and white American cultural values. Initially, Coke was developed as a healthy drink sold to middle-class Americans as a 'pick-me-up', an alternative to alcohol, with a potent mix of stimulants. The drink was then sold as something aspirational, for working-class Americans and for American children – if you wanted to be a successful 'go-getter' you drank Coke, if you wanted the girls to fall in love with you, you drank Coke. The drink became a fixture of American leisure lives, handed out at parties, bought at restaurants and bars, and sold to accompany sports events, holiday trips, cinema outings and evenings in watching television. Before and during the Second World War, Coke was seen by Europeans as something exotic and fashionable, and demand for the drink grew worldwide, driven by links to sports events and American power (Keys, 2004). By the 1970s, Coke had become a global brand, one of the most recognizable, its colours and logos adorning bars from South America to New Zealand. With the spread of coke comes coca-colanization, the spread of Western values of individualism and capitalism and the loss of cultural diversity: coke is at the vanguard of normalizing white, American popular culture and everyday leisure as universal human forms. With the fall of Communism in the 1990s in Eastern Europe the Coke brand became ubiquitous in the former Soviet states; and because Coke is non-alcoholic, it is an American brand that is welcomed in the heart of the Muslim world. Coke, the archetypical American drink, now sponsors soccer's World Cup. Coke has sold itself as a multicultural, post-colonial drink for the world, but its existence and its dominance of the market and popular culture are due to its white, American roots and the political and economic power of twentieth-century America (Ashbee, 2011). Like KFC and McDonald's, it is a brand that has made white American tastes to replace local tastes.

Shopping

Baudrillard's concept of hyper-reality can be used to understand the proliferation of shopping malls as a postmodern leisure experience (Hegarty, 2008). Shopping malls are a common feature of the Western world, though they are typically a feature of American life. The word *experience* is typical of postmodern accounts of the hyper-real – in postmodernism, all objective truths about ethics and aesthetics are reduced to the sensations of the body; the hyper-real is better than reality because it is better at stimulating the body's senses. Shopping malls represent this hyper-reality because they have replaced the high streets of towns and cities with bigger and more convenient retail outlets, housed under enormous roofs and stocked with a seemingly endless supply of fast food, coffee and sweets. Customers enter shopping malls and are overawed by the enormous scale, the bright lights, the colours and the images. Instead of shopping for the necessities of life, mall shoppers browse slowly, spending their free time following the signs and symbols of the mall to find more ways of consuming, more ways of spending. Retail in shopping malls is about fakery and creating false desires, but this is accepted by the consumers who willingly trade the reality of the high street for the hyper-reality of the mall. Shoppers drive to the mall, shop at the mall, eat at the mall, watch films and go bowling at the mall, and even stay overnight in the mall's hotel – so they can wake up refreshed and start again on the endless loop of bargain hunting. The mall has become a symbol of whiteness and Westernization in every country where the design has been replicated – even in the Arabian Gulf, shoppers eat donuts and popcorn and shop at Gap.

Buy Nothing Day is an annual, global, counter-cultural event held in November (O'Sullivan, 2003). Originally a campaign organized in Mexico by left-wing activists concerned about the increasing over-consumption of goods that were never needed, the Day was soon transferred to the United States. From there, it has become something celebrated in dozens of countries by anti-globalization campaigners, anti-capitalist activists, environmentalists, anarchists, pranksters, campaigners for local produce and local economies, left-wing activists, liberals and others concerned about the world's obsession with shopping. On Buy Nothing Day, individuals and groups pledge not to take part in shopping activities, a leisure pursuit they believe is a waste of resources, a drain on the planet and fundamentally unjust and immoral. Governments, transnational corporations, and transnational organizations such as the World Bank, support shopping and consumption as ways to increase Gross Domestic Product – a key marker for economic growth. For the anti-shopping campaigners, such consumption is unnecessary and damaging to the environment and to the individual, and the whole concept of growth, they say, is deluded on a planet with finite resources. Far better, say the campaigners, is to use time spent shopping doing some more meaningful and self-satisfying leisure activity, such as reading or walking. Campaigns such as this are deliberately provocative and are designed to make us think about what we are doing when we do buy stuff. It might feel satisfying buying the latest clothes or the latest electronic gadget, but this, say the anti-consumption activists, is an illusion. We are spending money we do not have on things we do not need. But the campaigns are Western campaigns, organized mainly by rich, white Westerners, who use their cultural capital to tell poorer people – white and non-white, Western and non-Western – not to do something they enjoy doing.

Reading

Books and other forms of literature were discussed in a previous chapter, but there is one aspect of reading relating to whiteness and everyday leisure that needs to be examined here: the tension between reading e-Book and reading paperbacks and hardbacks (for more on reading as leisure, see Stebbins, 2013). The technology of the e-Book now allows readers to buy and carry thousands of books in one device, and to read books in a way that does not feel too uncomfortable. The e-Book has the potential to make more people into readers across the globe – as the price of the technology comes down, more and more people will download books in the same way people download music. The e-Book also has the potential of opening up opportunities for the demographic of published authors to change: the publishing industry is dominated by white, Western elites, a closed circle of publishers, editors and authors who share similar backgrounds in private schools and elite universities. With the new technology, anyone can publish their book and sell it online. The rise of the e-Book has been swift and already by 2012 electronic downloads of books are out-stripping the sale of hard copies of books (paperbacks and hardbacks): bestsellers will increasingly be sold in an e-format, along with academic texts for students and possibly newspapers and magazines (Stebbins, 2013). The white, Western elites have already responded to the possible democratization of publishing and reading by shifting the discourse of authenticity. Paperbacks and hardbacks are now marketed and talked about as 'proper' cultural artefacts to be displayed on shelves in one's house, to be enjoyed and cherished, to be opened carefully, to be smelled. All sorts of *hoi polloi* read e-Books, all manner of un-educated and dim-witted authors publish e-Books. For the white elites, the 'proper' publishing industry preserves standards of style and standards of taste; for the publishing industry, this means increasing the price of books and working with High Street bookstores to move away from the mass-marketing of discounts. The higher the price, the classier the cover, and the whiter the reader.

Dancing

There is nothing more communicative than dancing to live music, especially if that music is being played free. Nightclubs offer people the opportunity to dance but such places have negative, instrumental connotations. They are commercial spaces designed to rob drunk and vulnerable people of their money through over-priced drinks and entry fees. The music played in nightclubs is not conducive to free expression, though it is possible to feel the thrill of physical movement in such places. Far better is a gig venue, where a band plays your favourite music. Musicians know the combinations of sounds that can move us, the heavy bass notes that tingle in our stomachs, the key changes that make us smile and wave our hands in the air, and those crunching metal riffs that make us bang our heads. In these liminal spaces, we can truly express ourselves, feel at one with others and find some meaning and purpose. Gigs are of course commercial transactions. People pay the musicians, they buy their records and T-shirts, and they make money for the venue and the promoter. So dancing only becomes truly communicative when we respond to live music that we haven't paid for – but that is a difficult thing to find. Once upon a time people sang and played music for themselves, while others danced to that music, but finding a public or private space where this happens is a rare thing in late modernity (Dudrah, 2011).

In the late modern West, dancing is a highly racialized form of everyday leisure. In the United States, for example, country and line-dancing are dances for lower-class white people, marginalized by their bourgeois white peers but still themselves members of the imagined white, American community (Counihan, 2008). Raves are superficially multicultural, but are dominated by middle-class white people, not just on the dance floor but especially the promoters and DJs who make money out of the scene. Nightclubs are stratified and marketed to various ethnic groups, classes and subcultures, with the expensive nightclubs in Western cities becoming exclusive sites of cultural conformity. Nightclubs have different forks of music and dancing on different nights – non-white music is often advertised as a special night during the week, whereas Saturday night is reserved for the white music scene (Buford May and Chaplin, 2008). Where white, Western everyday leisure intersects with non-white dancing practices, this is often an act of post-colonial expropriation by middle-class whites seeking the thrill of the exotic but in a sanitized, unproblematic form. So, for example, bellydance, particularly the modern Egyptian cabaret form, has been appropriated as a dance for white women to explore their femininity in a women-dominated space where postmodern feminist rules apply (Maira, 2008). The dance has spread across the West as a feminist leisure activity, but it carries with it the flavour of an Orientalism that ignores the reality of Egyptian popular culture and the struggles in Egyptian society. Similarly, salsa spread as a dance fashion at the end of the previous century for middle-class white people who wanted to show off to their white friends their appreciation of Cuba and Latin American culture – but it was a modified variant of salsa, and the young people of Cuba were dancing other styles, and listening to Western dance music and reggae at the time (Urquia, 2005).

Socializing online

When the Internet was formalized through the invention of the worldwide web system in the early 1990s, most early adopters were urban, rich, educated and liberal. These early users saw the Internet as a place of free discourse and communication, allowing them to bypass the systems of censorship, commodification and control associated with the entertainment and print industries (Briggs and Burke, 2009). Some early users even believed that the Internet would transcend nation-states to build a global, international community of free individuals, who would spend their time sharing ideas, spreading democracy and human rights, and creating a free space for artists, dreamers, philosophers and other utopians. This strong belief in the redeeming nature of the Internet as a free, uncensored site is still prevalent in discourse about the Internet, and on the net itself: Wikipedia, for example, would not exist without its creators giving their time and knowledge to edit the data. However, since the 1990s the Internet has become commodified, populated by multinational corporations wishing to sell to us directly (Amazon) or to turn our personal data into a commodity to sell to others (Facebook, Google). Our informal, private internet use is turned into browsing habits and preferences that are bought and sold by marketing companies (Briggs and Burke, 2009). Our time on the net is increasingly used to buy things. There are still millions of websites that give users free stuff, but even that transaction is one of commodity, buying something for nothing, rather interacting and contributing. Castells (1996) argues that the technological advances associated with the Internet have disrupted traditional social structures, economics and politics. In this information age, the Internet changes the way we consume popular culture, for example. We no longer listen to music on the radio or television and go out to buy it on the High Street – instead we browse file-sharing sites and swap recommendations with others. The Internet has made all culture popular and has bypassed traditional arbiters of taste, such as professional critics and corporations – there is an anarchic democracy of free choice. However, Castells does not believe that this makes the Internet an individualist utopia, a paradise of downloading and sharing. Rather, the Internet has become an apparatus of commerce, control and surveillance, with most of this activity hidden behind the discourse of personal freedom. The new corporations work alongside governments to balance the pursuit of profit with the need to limit political action. Popular culture on the Internet is increasingly marketized, with every transaction noted and tracked by algorithms so that the gatekeeping sites on the Internet can target adverts at us more effectively. At the same time, governments are keen to access personal data most of us have placed on social networking sites.

Communities of interest mark out the territory of the Internet (Papacharissi, 2011). To take part in any internet forum you are engaging with like-minded people, making assumptions about the things they might think, and taking with you a combination of knowledge and familiarity with the topic you are discussing. Social networking sites are models of community-building. On Facebook, users choose their friends by finding people they know in the real world, but soon they start to add friends who they only know through other friends. Then users join groups or become attached to other pages (the fan pages of a favourite band or sports team), and soon these users start to become friends with people they have only come into contact with in this online space. Some people see Facebook and other social-networking sites as a model of reality, something which is a crude reproduction of their real-world leisure interests but not a replacement. Others view these social-networking sites as something that is a replacement for real-world leisure interests. It is likely that both views of Internet communities are correct, depending on the individual and the communities in which they are situated – someone might arrange a walking trip with old college friends online, while at the same time following the activities of a famous celebrity. These leisured communities online are an extension of real-world communities, not the death of them – despite the scaremongering of tabloid newspapers users are able to move from online and real world at the flick of a switch, with no real dangers involved in confusing the real with fantastical.

The leisure activities and practices of subcultures are easily identified through watching television and movies, reading books and magazines, and surfing the Internet. If you want to become a 'petrol-head' driving souped-up cars, there is a magazine and a website to help you get started (Falconer and Kingham, 2007). You read the magazine to find out about where to buy a car to customize, what the best buys are for customizing your car when you've got and where to get the best deals. You also learn from the magazine the language of customizing and driving fast cars, the habits and practices, and the fashions. On the Internet, you can swap video clips of your car in action with other petrol-heads, you can boast about your own prowess, you can learn about identifying yourself with the romanticized history of the subculture, and of course you can learn how much to display and how much to hide. In all this activity, you may break some laws around driving safely and within the speed limits of your country, and you may join in the objectification of women and the outlaw posturing of your peers, but you do not leave the mainstream culture in which you live. Being part of the subculture of petrol-heads is no different to being in the subculture of lifestyle sports or extreme sports, such as snowboarding and surfing. The subculture is mediatized as part of popular culture: it is an artefact of the commercial world of the cultural industries, something that is packaged and presented to us as cool and edgy and alternative when it is far from it.

For a number of years in the first decade of the twenty-first century, Myspace.com played a significant role in the construction of modern subcultures. The business models of the pop music industry assumed young people were passive consumers who could be sold one artist after another in a charts system controlled by marketing teams. Big labels made big profits while the labels controlled the supply of information about bands and the availability of music in record stores. All this changed with the creation of Myspace.com, combined with the advance in technology that led to the rise in music-file downloading (legally and illegally) as an informal leisure activity. Bands could now market themselves directly to potential fans without the need for managers or labels; they could sell their music; and fans could find bands they liked through following hyperlinks from one band's page to another band's page. Myspace.com became something that shaped modern pop music subcultures, making underground and extreme music genres and artists freely available and easily accessible to a global audience (Wilkinson and Thelwall, 2010). Some pop music subcultures were invented through Myspace.com's friend connections (such as deathcore in heavy metal, based on the hype of counting thousands of friends). Myspace.com was bought by a transnational corporation seeking to profit from the hype. The pop music industry saw Myspace.com as something to be exploited, playing with hyping and friend-making to try to build fan bases for bands, using online fan networks as they used real world street teams to hype acts through word-of-mouth. Subcultures drifted away from using Myspace.com as soon as this started to happen.

It is in conversations and networking online that hegemonic whiteness operates, turning everyday leisure into a form of instrumentality. White people normalize their particular leisure interests as something universal and the leisure interests of non-white people are marginalized, ring-fenced or mocked or fetishized as something irredeemably Other. The Internet allows anyone to find their community and belonging in some website or social media network. But the power of the white West means there is more economic, social and cultural capital available for white people to dominate the 'neutral', 'mainstream', de(re)-racialized popular hubs. The most popular social network sites allow white users to create social networks that avoid non-white people and non-white interests, and create a sense of imagined and imaginary community that makes such whitewashing a barely conscious act. In turn, the economic and hegemonic power of the white, Western entertainment and media industries limit the choices available in their online spaces to ones that replicate middle-class, white, Western tastes and preferences. Discussion on the Internet is often in the form of angry objections and irrationally held opinions, and sexism and racism are rife where the social rule of anonymity fosters the culture of trolls (Papacharissi, 2011). But even where conversations are polite a code of white norms and values is often implicitly operating to guide discussions from questions of inequality and power to the myths of individualism and meritocracy that keep the white people with power in power today.

Ethnic food

Eating out is an everyday leisure activity that transcends class in the West, and is something that is a leisure activity in many other parts of the globe. In the West, eating out is a product of two competing social trends: the elite construction of fine dining and distinction through food snobbery, and the industrialization of food production, which has allowed processed food to be produced in great quantities, stored, shipped from place to place and sold relatively cheaply. Both the mass market takeaway/café and the up-market restaurant give their customers a space in which they can eat, drink and socialize, while paying for the privilege of eating food produced by others. For the elite white, Western classes, restaurants are places of unwritten rituals and hierarchies, where elite whites can show off their high culture and their place in society and be served by lowly waiters (often non-white or poor, working-class white people). For the white working classes of the West, eating out is often a way of sustaining high modern, white working-class traditions, constructing a sense of national pride and whiteness through the eating of supposedly authentic foods such as fish and chips in England, or smoked sausages in Germany (Olsen, Warde and Martens, 2000).

   Ethnic food is a commonplace term used in everyday white, Western culture and writing about the food industry. It refers to styles of food – particular foods, particular flavourings, particular cooking styles, particular modes of preparation and presentation, and particular modes of eating – that supposedly represent an authentic culinary tradition from some country (or region) foreign to the country in which the white person using the term lives. In the United States, ethnic food is Mexican, or Japanese, or African. In the United Kingdom, ethnic food is Chinese, or Indian. In France, ethnic food is Algerian, or African, or Vietnamese. The makers of the food might actually be non-white minority ethnic groups long-established in the country, people who are culturally and politically citizens of the country in which they have been born – but whose food is pigeon-holed into the box marked 'ethnic'. All food that supposedly represents an authentic tradition from a nation-state should be called ethnic food (it could be argued, of course, that all food has roots in the traditions of one ethnic group) but the logic of ethnic food is flawed. 'Fish-and-chips' is not sold as the ethnic food of the white British majority ethnic group in Great Britain – it is 'just' fish-and-chips. The majority ethnic groups in Western countries make their foods the normal foods of eating out, or they invent traditions that make modern foodstuffs normal bit invisible white culinary habits. By labelling foods associated with the cultures of non-white minority ethnic groups as ethnic, what is white becomes non-ethnic, that is, what is white becomes normal – and the non-white becomes exotic and foreign. The hegemony of whiteness turns hybridity and change in the food eaten by people into fixed, unchanging ethnic categories associated with outsiders and foreigners. This might mean some such foods become embraced by the majority white communities of modern Western nation-states – such as pizza or the taco – but such adoptions change the foods entirely, and are always on the terms of the majority white culture: subject to fickle fashions and appropriation by celebrity chefs, opinion formers and corporations.

Ethnic food is a product of the industrialization, socialization and marketization of eating (and eating out) and everyday leisure in late modernity, but it has its roots in an earlier period of modernity: the age of Empires and protectionism in the nineteenth century. Curry in the United Kingdom is a good example of the early adoption of the idea of ethnic food in the marketized restaurants of high modernity. The British Empire's slow acquisition of India from the eighteenth century onwards saw British economic and political power shift from the Atlantic to the Indian Ocean. Thousands of Imperial adventurers working for the East India Company and the British Government – soldiers, merchants, administrators, sailors – plus equal numbers of private explorers and exploiters, served time in India. Some became successful entrepreneurs, importing and exporting goods and fashions. Cloths from British mills found a large market in India, but equally, Indian products such as tea and spices found a market in Britain. The curry, a dish served with a sauce flavoured with Indian spices, became a part of British eating habits in the eighteenth century (when there is the evidence of a curry recipe in a cookbook), and by the early nineteenth century the first curry restaurant had opened in London (Sen, 2009). It could be argued that curry houses have an older pedigree than fish-and-chip shops (which only developed in their existing form in the middle of the nineteenth century, when new frying technologies made the food possible), although the first curry houses in the early nineteenth century had a transitory existence relying on white imperialists and Indian sailors for their custom. Whatever the history of particular restaurants, the curry became a popular and a bourgeois meal, available in cheap curry houses and expensive hotel restaurants, by the end of the nineteenth century, its popularity due to the popularity of Eastern fashions, the Orientalism of the white Imperialists and the appreciation of the taste by old colonialists.

Into the twentieth century Indian curries declined in popularity from the Victorian era, possibly as a result of the decline of the Empire and the rise of Indian nationalism. However, curry houses continued to exist in areas where Indian sailors (mainly from what is now the Sylheti region of Bangladesh) had settled, such as the East End of London – and Indians occasionally opened up curry restaurants aimed at the elite whites of Britain. Curry grew in popularity as a result of the end of the Empire and the end of the Second World War – a time when British industries and shipping companies were so short of labour that they purposefully advertised and encouraged inward migration to Britain from former colonies. Indian, Pakistani and Bengali (Bangladeshi) men fleeing poverty and violence, or seeking opportunities denied them back home, came to work in the former colonial power – and like all new migrants, they were forced through racism and prejudice to live together in the poorest parts of the towns and cities (Buettner, 2009). Many of the men came from the same area: the Sylhetis who worked in shipping; and the Mirpuris who worked in the mills of Bradford and East Lancashire. In the 1950s, some of these men opened up curry houses mainly for the use of other migrants – in Bradford, for example, the Karachi and the Kashmir were opened by 1958, the former in a terraced house, the latter in an old shop near the centre of the city. Although they were opened to serve as social eating venues for migrant workers, these modern curry houses soon became known to white British folk looking to eat out at a cheap price. The curry houses in Bradford were replicated in the Sylheti eating establishments on Brick Lane in London, where most of the Bangladeshi migrants lived; again, places opened up to serve migrants attracted white British people, who had little choice in the 1950s when it came to eating out at a cheap price (Ray and Srinivas, 2012).

By the 1960s, this first generation of migrant men from South Asia had been joined by women and had started to make permanent homes in the United Kingdom. People married, had families, moved out of the industrial or shipping work that brought them to the country and found jobs elsewhere. These migrants struggled to be accepted in the places in which they settled, but there was inevitably both integration and hybridity. They had children, gained British citizenship and became entrepreneurs: many Sylhetis moved around the country to open up 'Indian' curry houses, so that the modern, twenty-first century 'Indian' takeaway and restaurant trade is dominated by Bangladeshi cooking styles and British Bangladeshi staff (Buettner, 2009). As the traditional industries the first generation of migrants had come to work in collapsed; their communities were hit by unemployment and poverty, which led to entrenchment in inner-city areas, but the communities were resilient enough to survive and prosper as new Britons (even if many of the older, white British people, the politicians of the right and the right-wing newspapers, attacked migrants and migrant communities and accused them of not being 'proper' British – a fear of the Other that resurfaced as Islamophobia in the months and years following 9/11 – see Sheridan, 2006).

For white British people, the curry has become a national dish, with the chicken tikka masala, reputedly invented by a British Asian curry chef, becoming more popular than supposedly traditional dishes such as fish and chips. Even the leader of the far-right British National Party, Nick Griffin, has admitted he likes going out for a curry. By the 1970s, going for a curry became a ritual post-drinking experience in small towns up and down the country, where the aim was to drink cold lager and eat hotly spiced curries such as the Madras and the Vindaloo. In such eating experiences, there was of course a tension between the mainly white customers and the mainly Asian, Muslim staff. People who go for a curry after an evening drinking alcohol can be awkward, aggressive and uninhibited – racial abuse of the staff used to be common and still occurs, along with fights between customers and staff. The racial politics of Britain – the poverty of British Asians, the powerlessness of working-class whites, the fear of the Other and, latterly, the fear of Islam – have all shaped the encounters in curry houses since the 1970s. However, for many white people, going for a curry is something they now do at any time, not just when the pubs close: people go for a curry for lunch and in the evening for dinner; they go with friends, family or work colleagues; sometimes they drink and sometimes, in curry houses that do not have a licence for alcohol, they do not drink; they go to cheap places with formica table and get a curry with rice or chapattis; they go to exclusive restaurants that promise fine dining experiences

and fusion cooking; they order the traditional British curries shaped
by Sylheti cuisine but invented for white British palettes, such as the
Korma, the Bhuna, the Vindaloo; they seek out restaurants that serve
authentic Indian cuisine; they make curries at home.

*It's a treat. It's good value. It's so much better than the bland English
food you used to get. It soaks up the booze. It's one of the best things
about this country. It's the only places to eat in my town that isn't a Pizza
Express… White people I know all go for a curry. I don't know any white person of my age or younger who doesn't like curry. As one person said to me –
the only people who don't like it are old folk who aren't used to the spices. But
even the older white British generation has grown up eating curry, and now it's
a commonplace dish served in retirement homes. White people even express
nuanced preferences about the different regional styles available – as someone
Yorkshire-born I prefer the heavier, drier curries of Bradford to the wet curries
of London and the south of England, or the strangely coloured, sweeter curries
of Scotland.*

The curry, then, represents the continuing power of whiteness in
post-colonial Britain. There is agency at work in the British Asian communities that dominate the curry industry: successful businesses make
their owners respectable locally, especially if the businesses become
involved in charity work; and the industry offers British Asians a way
of interacting with and re-shaping British culture into something more
postmodern, hybrid, multicultural. There are jobs available for British
Asian men (and, gradually in some places, for British Asian women)
as waiters and kitchen staff, which in the post-industrial economy of
Britain might be the only job opportunities available. But the industry
and the everyday leisure experience depends on the whims of white
people and the policy-making apparatus controlled mainly by white
politicians. Opening up a new takeaway can be a traumatic process
when local people object and planning officers and councillors take
notice of those objections – there are also licensing and regulatory matters that can work to stop a curry house from opening. Indian food
is fashionable now but fashions change – in the 1970s white people
preferred to eat Chinese food. Where Indian restaurants have become
successful (and where they have turned into chains) is at the top end
of the market, where they cater for bourgeois white people seeking a
measure of authenticity and the thrill of diversity, where white people can show they are at ease with the Other and impress their white
friends with knowledge of 'how the Indians eat'. Indian restaurants for
British Asian people and for a more diverse range of the market do
exist in multicultural cities, but these are outnumbered by the high-end

restaurants and the cheap Sylheti places that serve as the only places in many small English towns where white people see Asians. Ethnic food, then, is highly problematical as a concept, and even where it has been in part an invention of minority ethnic groups (such as British curry) it is still subordinate to the hegemonic forces of whiteness that shape contemporary Western culture.

Conclusions

Everyday leisure activities should be forms of leisure that are mainly communicative in nature. However, even these activities are shaped and limited by instrumentality. As such, they become ways in which instrumental whiteness is constructed and made invisible: whiteness itself becomes the everyday normality of Western life.

10
Whiteness and Tourism

Introduction

We are away on holiday on the Hebridean island of Skye. Back in Scotland again. Last year we went to the south-west of Scotland on a pilgrimage to see the locations where the classic horror film The Wicker Man *was filmed. We had printed pages from a website to guide us to the pub that served as the interior of the pub in the film, and the old offices that served as the exterior (in another village completely – the magic of movies!). The film is now a cult thing, worshipped by a few dedicated fans and modern pagans, appreciated by a wider, knowing, liberal bourgeois audience – but there wasn't a tourist trail to guide us. When we went looking for the site on the cliffs where the wicker man was built (where the climax of the film sees the good Christian police sergeant Howie meet his fate at the hands of Christopher Lee's pagan lord Summerisle), all we knew was it was somewhere beyond a caravan park, overlooking the sea. We wandered up and down the edge of the park in the rain, desperately checking the wet pages in my pocket against the misty cliff edge. Meanwhile, white, working-class families in the caravans stared at us suspiciously as we disturbed their breakfasts.*

Back in Scotland again. Our second trip to Skye. We're in a cottage, doing the white, middle-class staycation thing. Don't go abroad, stay at home in the United Kingdom, but find a place that isn't over-run by tourists, go somewhere real, somewhere authentic. Go somewhere with hills and deserted beaches, see seals and dolphins, go walking, feel relaxed away from the stresses of modern, urban life. The white, middle-class people I've spoken to for this book all book cottages around the United Kingdom, once or twice a year (some go abroad as well, they have children who want to see the sun). We all share our tips about places we like, villages and towns and areas that we go to because we think they are the perfect escape, off the beaten track, away from the masses.

*Of course our snobbery and seemingly good taste and our concern with finding
the real Scotland (or Wales, or Cornwall) is a lie – we are all following other
people like us up the motorway to Scotland or wherever it is we are going.
We are bourgeois white folk who follow the white, middle-class fashion of
finding a cottage in the middle of nowhere – on holiday in Skye, in our cottage
booked online, I can see dozens of cars pass by on the main road across the
bay. Half of them are white folk from the European mainland, who tend to be
younger couples or small groups. The other half is middle-class white English.
Now I know some of them are here because they love the outdoors or the
wildlife. But I know some of them are here because the urban life they are
escaping is a multicultural life: all these cottagey places white, middle-class
people go to are the whitest places in the United Kingdom. Can that be a
coincidence?*

This chapter will focus on three areas of tourism where there is
existing research to enable a detailed secondary analysis: package holidays in Europe and the Far East; heritage tourism in the United
States and the United Kingdom; and independent travelling along
well-used routes from the West (and Australia) to the East and South
America. The chapter will begin with a short overview of the history
of the modern tourist industry and the assumptions about whiteness
and (post)imperialism implicit in the concept of travelling to another
place 'on holiday'. The section on package holidays will discuss subordinate whitenesses based on working-class Western identities and
identify a residual cultural resistance to the norms of instrumental
hegemonic, white, middle-class Westernness. The section on heritage
tourism will discuss the ways in which dominant whitenesses are imagined (including instrumental whiteness), and the instrumental nature
of that (re)imagining, which privileges elitist narratives of place and
nation. The final section on the subcultural identity of the traveller
will explore the tensions between traveller beliefs about cultural diversity and the white neo-imperialism and colonialism such travellers bring
with them to the countries they visit.

History of the modern tourism industry – A brief overview

Most cultures have holidays, festivals, special places away from home
and pilgrimages of varying intensities. The great Muslim pilgrimage
of the hajj has deep roots in Islamic culture, and it inspired both the
creation of a pre-modern tourist industry around it and a similar pilgrim experience and industry for medieval Christians (Spracklen, 2011).
Many pre-modern cultures created nascent secular tourism for their

elites – the Romans ruling classes, for example, visited Greece to see the sights following the guidebooks of the day written by Pausanias and others (Spracklen, 2011). It is this latter example of secular touring that inspired the growth of the Grand Tour, the long vacation to Rome and other centres of culture and civilization undertaken by elite white Europeans in the early modern period. But modern tourism, while drawing on the pre-modern ideas of the pilgrimage and the public demonstration of elite cultural capital, only really established itself in the form in which it is familiar today with the advent of flows of technological power and economic power. The former power was the power of the steam engine and the telegraph, systems that enabled scientists and engineers to bring distant places much closer – physically and psychologically – than they had ever been. Railways and steamships spread across the world from the West and allowed Western nations and corporations to win empires. This flow of economic power, the second flow that gave rise to modern tourism, led to the increased bourgeoisfication of the white West, increasing the numbers of the rich and well-off, the new money that aspired to spend their cash in acts of conspicuous consumption. Fuelling the profits of the bourgeois classes were the white, working classes who lived in the booming towns and cities. They may have been exploited but they still earned more than peasants working in rural areas, and thus they had a small amount of money of their own to spend on leisure activities.

In the new capitalist economies of the West in the nineteenth century, entrepreneurs and companies had the capital and the freedom to establish new leisure activities for rich and poor alike. Tourism for the elites was fairly well established in the late eighteenth and early nineteenth centuries, but this grew with the increasing numbers of the middle classes. Tourism for the working classes had to wait for legislation that gave workers time off at the weekend. Once the working classes had established the right to have holidays, it became common to see entire towns in the United Kingdom shutting down their factories for a week so that their workers' holiday clubs could organize days away to the resorts on the coast. This pattern was repeated across the West. The modern tourist industry made money first from the ruling classes, then the bourgeoisie that emulated the ruling class, then the working classes who were encouraged to spend their money freely in their leisure time. Holidays for the working classes were often paid for and organized through subscriptions to friendly societies or voluntary associations – gradually, in the twentieth century, such associations decreased in number as the transnational corporations founded in the nineteenth century took over the market.

Tourism and global travel mark the flows of ideas and power identified in globalization theory (Gonzalez, 2010). Initially, tourism was a leisure pursuit for elite whites, but industrialization and capitalism in the nineteenth century opened up tourism to working-class Europeans. As technological flows shortened the distances between countries, foreign travel became an accepted part of normal tourism. Elite Westerners, in response, demonstrated their power and status by travelling beyond the West to far-off, exotic destinations: the Alps to ski; Monte Carlo to gamble; and so on. The power of the elites transformed these destinations into colonies of the Western ruling classes, a transformation that continues today. At the same time, new elites have emerged in Russia, China and India, who use their wealth to visit the West. These travellers are welcome in Western countries that have lost some of their power in the post-colonial age: places in the United Kingdom such as Bath, York and London rely to an extent on these elites spending their tourist pound there. However, the free flow of non-white travellers at the top end of the power scale is not mirrored at lower levels of the power scale: poor migrants looking to find temporary or permanent work in the West are discouraged from taking advantage of globalization's smaller horizons, and border controls continue to stop such migrant flows. It is okay for the elites to travel across the globe, on vacation or to find work opportunities, but governments close the doors on others wanting to do the same.

The quest for authenticity in tourism has been the subject of hundreds of research appears since MacCannell (1973) first described the way in which tourists seek some true, authentic experience of community in the places they visit. Modern life according to MacCannell had become too fake, too inauthentic, for educated, white, Western tourists. They could not find community and belonging in their work or in their cities, but they believed they could find community somewhere else. This notion say many tourism researchers still holds true today, years since MacCannell's first paper on the subject. Tourists want something authentic because they have no authentic community back home. They believe that travelling abroad will bring them to some community where such authenticity still exists – they are travelling in pursuit of something more real, something more 'true' than the plastic world of the West. Tourists want to find the place where the locals eat out with their friends and families; they are not satisfied with the restaurant in the hotel. Tourists want to get off the beaten track, to live like the natives, to experience the sense of belonging and well-being that comes with this authentic community. Of course, there is no such thing as authenticity – every community is created with tensions, debates over the boundaries, paradoxes of meaning and fragile roots – but just because no community is authentic does not stop authenticity being of interest to researchers interested in the motivations of travellers. The very idea of authenticity is an essentialist one, based on a Social Darwinist idea of racial purity. Foreign, non-white cultures are authentic because they are closer to a pristine state of nature – they are more authentic because they are not as civilized or as advanced as the white tourists; they are objects of pity as well as envy, and the white gaze fixes them in their supposedly pristine state for the West's delectation. The white West is inauthentic because it has lost its 'roots'; it has (it is implied) become hybridized: the logical extension of the argument is that inauthenticity in the West is based on a racialized fear of miscegenation and the decline of 'pure' whiteness (and white hegemonic power).

Package holidays

Modern tourism is the mass-market industry that emerged in the middle of the twentieth century, which catered for all sections of society. In the United Kingdom, the mass-market tourist industry appeared earlier in history, in the nineteenth century. This was due to the change in labour laws that gave workers more free time and more power to negotiate holidays. The boom in the white, working-class tourist led to the boom in working-class tourist destination resorts such as Blackpool, Margate and Scarborough. These tourist resorts offered sand, fish and chips, rock (candy sticks made of hard sugar) and the opportunity to have a few drinks and a stroll along the promenades. Because these resorts and tourist experiences occurred before the mass migration in the second half of the twentieth century, these tourist vacations were typically representative of the wider high-modern, imperialist white British culture. Blackpool became infamous for its reputation as a place to go to get drunk and have sex with strangers, though the entrepreneurs and landlords/ladies who ran the hotels and guest houses did their best to project (and police) a more sedate image of the resort. In Blackpool, the white capitalist mill-owners of the North of England could find sedate pleasures such as walking in pleasure parks themed with Oriental or Exotic designs, or going to see exhibitions of colonial art. For their workers, comedians offered music-hall jokes about the stupidity of the Irish, the laziness of black people and the untrustworthiness of Orientals and Jews (Weaver, 2011). In the circuses and in the carnival shows, black and other non-white people were paraded as freaks of nature, or primitive savages: or as exotic flukes with magical powers (Parker, 2011). After the genocide of the Holocaust, much of the anti-Semitic nature of the end-of-the-pier shows subsided, but the racism aimed at black people, Asians and foreigners continued to be the subject of comedy performances into the last quarter of the twentieth century. As the minority ethnic community grew in Great Britain, the old white, working-class resorts became places of retreat for white people – liminal spaces where they could go to see club comedians such as Bernard Manning telling sexist and racist jokes; where they could be sure there were no black people; where they could holiday and retire as if the British Empire still existed (Weaver, 2011). For the bourgeois white classes, the Cornish Riviera, Bournemouth and Eastbourne allowed them to live the same racially exclusive dream – places where they could live the lives of colonial officers and bureaucrats, drinking tea and gin and tonic, playing golf and being served by silent, subservient hotel staff.

The mass market in the United Kingdom was replicated across Europe and the United States – and as modernity spread across the world, mass-market tourism followed. Once technological advances had made overseas travel feasible for the working-class tourist, package holidays became the norm. The working-class tourist from England flew to Benidorm in Spain, for example, where cheap hotels were built close to the beach and each other, and close to the strip of bars named after places in the United Kingdom (Obrador, 2011), and the Germans preferred to fly to the Balearic Islands (though of course all the white, northern Europeans flew everywhere seeking the sun: Greece, Italy, Florida and Turkey). In Benidorm, the white working-class British can see blue sky, get a tan, sit on the beach and swim in the sea – all things to be appreciated after living through a miserable grey, wet British winter. But they can do more: they can eat fish and chips, play golf, enjoy afternoon tea and ballroom dancing, shop at Marks and Spencer, drink British lager and bitter, and spend every day of the holiday being surrounded by people like themselves. The companies that sell package holidays understand their markets and their products. Their brochures and websites offer young, single people specialist holidays where they can get drunk and have sex. They sell destinations and hotels aimed at the bourgeois white middle classes, who want to play golf and be pampered in exotic places where there won't be any young people disturbing them. And they sell family holidays to the struggling working classes who feel the need to prove to their friends that they can afford to go abroad, that they are respectable and upwardly mobile. In Western countries where there is only a small minority ethnic community, the marketing material of package holidays is dominated by the smiling faces of white people, with only the occasional token non-white face or family (Aitchison, 2001; Edensor, 2002) – the implication is the package holiday will take white people to a place, which though in a foreign country, will still be essentially white. In the United States and other Western countries where there are significant minority ethnic communities, with growing minority ethnic middle classes, the tourist industry has become more willing to embrace diversity in its marketing and in its planning – but the industry is also aware of the need to pander to the holiday desires of the majority white community, which even in the United States is marked by a desire to have a vacation in some place where they can be surrounded by other white folk (Hargrove, 2009). Cruise ships are a perfect example of the hegemonic power of whiteness in the West: older, bourgeois white folk get on the ship thinking they are replicating the white upper-class world of the early twentieth century, when only the rich could travel abroad for a holiday (Weaver, 2011). White folk on cruise ships go from Miami or Southampton or Hamburg to exotic locations, where they are treated to entertainment by the local natives. They get served by non-white waiters in the ship's restaurants, play cards and deck games, go on trips to see the sights and buy trinkets off the smiling foreign non-white people in the markets.

One of the biggest companies – and certainly one of the most (in)famous – in the tourist industry is Disney. Along with their media interests and their global chain of stores, they own a huge estate of entertainment sites, hotels and apartment complexes and target families with young children. The original Disneyland theme park and resort opened in California, designed by the company to make profit from the loyalty of the children and adults who had fallen in love with the cartoons. In Disneyland the company maintained a strict, all-American, middle-class world, where workers had to follow strict rules about their appearance to fit the stereotypes of the high school: this meant Disneyland was exclusively white, designed to a white template, creating a whitewashed version of the American dream (Bryman, 2004; Wallace, 1985). Disney replicated Disneyland, creating enormous resort complexes in Florida and theme parks in Tokyo and in Paris. The idea of going to Disney for a vacation has been globalized, at least in the Western world: European children of all ethnic groups clamour to go to

Orlando to see Mickey Mouse, and the company has publicly distanced itself from the racialized myth of America it sold in the 1950s and 1960s (Breaux, 2010; Bryman, 2004). However, the Disney concept remains exclusive and wedded to a worldview and a global politics that promotes white American popular culture as a universal human good (as I have discussed earlier in this book). Going on holiday to Disney World in Florida is a way for white Americans to avoid having to go abroad, a way for them to experience the exotic in the form of Disneyfied representations of the Other. For white Europeans, going to Disney World is a way of going on a foreign holiday that feels safe and comfortable: the natives are mainly white, they speak English and the local customs are familiar through a hundred years of Americanization of Europe. The economic power of white people ensures that the Disney dream-world retains a smell of whiteness about it.

The whiteness of package holidays can be seen more clearly in the growth of long-haul vacations. By the end of the previous century, the increasing size and wealth of the white middle-class from America, Europe and Australia created a market of older couples (often without any child-raising responsibilities) looking to visit exotic, far-away places – but who were reluctant to organize the vacations (the flights, the transfers, the accommodation) for themselves. These people looked at package holidays in the short-haul destinations as being too working class, too downmarket. They wanted to go on holiday somewhere that priced itself out of the range of the lower classes they feared and resented – so upmarket, long-haul destinations started to be sold by the package holiday companies to this kind of customer. Places such as South-East Asia and West Africa started to build exclusive resorts, hotels protected by massive walls and fences, hotels that offered holiday-makers every possible comfort. Holiday-makers could stay in the resort complexes for the entire length of their vacation without going outside, because everything they needed was inside or brought inside to show them. For white Westerners with conservative, bourgeois inclinations, this type of package holiday was perfect: they could go somewhere exotic where they could take pictures of elephants and foreign dancers, so they could show off to their golfing friends back home; but they were safe and did not have to mix with the locals, who they suspected to be thieves, or eat their food, which they thought was guaranteed to give you stomach trouble. In the global recession such holidays continue, but there has been a reduction in the size of this part of the tourist industry – the poor countries trying to earn the white, Western dollar by building massive hotels now struggle to fill them (Roessingh and Duijnhoven,

2005). In other words, the economic problems of the rich, white West have had a dangerous impact on the well-being of those countries that made their own economies dependent on this colonialist relationship.

Heritage tourism

Heritage tourism is an important part of people's leisure lives in modern, Western societies – whether the Wild West towns of the United States or the old forum in Rome. Many people who participate in visiting heritage sites will encounter re-enactors, professionals or more likely amateurs who dress up as authentic people from a particular historical culture (Carnegie and Mccabe, 2008). In the United Kingdom, there are strong re-enactment scenes around the English Civil Wars, organized by the members of the Sealed Knot Society. Members of the Sealed Knot join real regiments from either side (Royalist or Parliamentarian), and act out battles and skirmishes from the wars, kitted out in reproduction armour, clothes and weaponry. But the re-enactors don't just act out the fighting: before and afterwards, they live in camps in character, addressing each other where possible in stylized versions of the English language of the seventeenth century, and maintaining seventeenth-century notions of masculinity and femininity (women as camp followers, not soldiers, a stance that some re-enactment scenes have abandoned). People re-enact for all kinds of reasons, but the main ones seem to be part of a living his-torical subculture, and to find community with like-minded individuals. There are subcultures of these subcultures who insist on authenticity in everything, including the make of tents and the use of toilets and showers – and others who insist on keeping hold of the luxuries of modern life. Of course, for the re-enactors who do want to be seen to be authentic, there is a need to keep the re-enactments as places for white people to play at being white people *in the past*: like the debate among film critics and fans about hiring black actors to play roles that were once considered white (for example, Heathcliff in a recent adaptation of *Wuthering Heights*), re-enactment groups and tourists vis-iting re-enactments are faced with how authentic they make the spaces. Other important re-enactment scenes are the Civil War enthusiasts of the United States: here, though, the racist values of the slave-owning South are far more problematic for these re-enactors. The fact that all re-enactment is a present-centred construction of the past, a problem-atically subjective (re)creation of something we can only ever know through interpretations, sparse primary sources and artefacts, might be a reason to de-racialize the performances and to allow non-white actors to play European settlers of America or Viking raiders in Denmark. However, most re-enactments ignore their own construction as present-day artefacts and attempt to be 'racially accurate' in their presentations and performances. So in the 1940s and Second World War re-enactment community in the United Kingdom, it is quite possible for black and Asian people to dress up and join in, but they become soldiers of the Empire or General Infantrymen (GIs) from the United States.

Some re-enactment and heritage tourism associated with re-enactment is inclusive and counter-hegemonic in its politics (the industry that tells the story of slavery and the genocide of Native Americans, for example). Then there is the obviously suspect politics of American or British Second World War re-enactors who dress up as German soldiers and fetishize the Third Reich (de Groot, 2006). But much of this form of heritage tourism is used to perpetuate less overt notions of national and racial essentialism, to make a case that the existing white, Western hegemony is one based on well-established roots that are exclusive to only those who have blood links with the past society that is being re-created. Tourists want to see re-enactors because they want to confirm their own belonging, their own connection to the past – re-enactment tourism becomes a form of nationalism and cultural chauvinism, where white people uncomfortable with the multicultural reality of modern Western life can find a more white space to visit. Re-enactors themselves could be described as tourists visiting and creating their own past, becoming concerned with living a life that is more real than the urban late modern lives from which they retreat. The myth of the idyllic, feudal pre-modern past has a strong grip on contemporary Western nationalism, and re-enactment tourism is just one obvious way in which that myth is expressed and (re)produced. The past in the United States, for example, becomes a past where America is discovered in 1492 by white Europeans, where white Pilgrim Fathers settle and give thanks, where white colonists rebel and make the country, where the nineteenth century is the story of Civil War (and the generous gesture the white president makes to the black slaves) and white cowboys.

In many Western countries, where communities based around one traditional industry such as mining have been replaced by commuter towns, shopping malls and post-industrial decay, the remains of such industry have become contested heritage tourism sites. Many post-industrial regions see heritage tourism as a way of regenerating their struggling towns and cities. There are precedents for such regeneration working as a result of turning old mills and mines into industrial heritage attractions. Tourists want to visit heritage sites and embrace history, whether it is the history of settler colonies in America and Australia, or the stately homes that recreate an idealized Edwardian past in England, or the battlefields of France where you can stand silent amidst the graves of some of the millions who died in the First World War. Ex-industrial sites are one other form of heritage tourism, and as such they cater for those people who want to feel they have visited a destination, or an attraction: steam engines are polished and pump away happily, engineers fiddle with equipment and mannequins dressed in period costume bend to operate a spinning jenny. Trade unionists and other working-class activists have campaigned for alternative histories to be foregrounded at these industrial sites: the strikes and struggles for better pay, the deaths at work and the callous owners of the factories. Some of this working-class history is now visible at industrial heritage sites, as curators and directors understand the nature of the industrial past and the struggles associated with the people who worked in the mines, mills and factories (Rudd and Davis, 1998).

Heritage tourism is problematic because it relies on a nostalgic sense of history and belonging, of community, identity and nation. In the summer of 2011, the streets of London and many other cities and towns in England were the sites of unrest, in which multi-ethnic, young working-class Englanders looted their local shopping centres for cheap trainers, wide-screen televisions and cigarettes and alcohol. At the same time in the countryside of England, the members of the National Trust – the multi-million member charity set up to own, manage and preserve built heritage and landscapes in England and Wales – continued to visit stately homes, where they admired antique furniture and ate cream teas (Jenkins and James, 1994). The National Trust was initially set up at the end of the nineteenth century to preserve English and Welsh landscapes from unwanted development and to protect what we might now call the natural environment of a place. It soon turned its attention to buying or acquiring stately homes, massive houses in private parks, from their owners, especially in the decades following the First World War, when many landowners struggled in the decline of the Empire. These stately homes were built on a huge scale from the eighteenth century onwards, but by the twentieth century many of them were demolished or in disrepair – they represented the economic power of the British Empire and the enormous profits made from the exploitation of others. The National Trust often leased stately homes donated to them back to the owners who donated them in the first instance – even now, many of the Trust's properties have aristocratic tenants. By the beginning of the twenty-first century, the National Trust had established a huge property and land portfolio, and a mass membership of millions – putting it in the top rank of landowners in England and Wales, and making it one of the biggest voluntary/third-sector organizations in the United Kingdom (with a membership that is bigger than that of the Conservative or Labour Parties, the two main political parties in the country). The National Trust serves as an interesting case study for the wider theoretical discussion about the role of heritage leisure in preserving and reconstructing elitist social identities and power relationships in these turbulent times of globalization, commodification and neo-liberalism (Roberts, 2011). The National Trust is known primarily for its stately homes which appear in period dramas on television and in the movies, and which serve as aesthetically pleasing images for the front pages of dozens of local and national tourist brochures: they represent a particularly elite white Englishness recognized and desired by domestic and foreign tourists. The Trust owns other, smaller properties, along with working farms, holiday cottages and millions of acres of countryside – but it is the stately homes that take up most space in the National Trust Members' Handbook, and it is these homes that make the most income for the Trust from visitors (members and non-members alike can visit the stately homes, though only members get free entry – and both members and non-members alike pay for the extras such as special guided tours, guidebooks and tea and cakes in the tearoom).

In my own experience of visiting National Trust sites, there is some difference between the tourists who visit the stately homes and those who walk in the landscape spaces, but both sets remain predominantly white and middle class. In the stately homes, older white folk and bourgeois white families shuffle through carefully conserved grand dining rooms and bedrooms, nodding at the strangeness of the white British aristocracy, imagining themselves as lords and ladies of the manor. They gasp at the kitchens and other rooms where dozens of servants worked. They rarely ask how these rich people earned their money or think about the exploitation of the poor and the inhabitants of far-off countries. The people who visit the National Trust's landscape spaces are lovers of nature and the outdoors, people who like exercise and outdoor pursuits, whether walking a couple of miles to a picnic spot or hiking 27 miles along the cliff-tops. In its newsletter, the National Trust includes adverts and publicity material for products that companies think will be appreciated by the Trust's members. These products are typically expensive pills for digestive problems, comfy shoes and slippers, Royal Jubilee clocks and plates, cruise ship holidays, steam train adventures and so on: products aimed at older, richer, white people, those who read the mid-market tabloids such as the *Daily Mail* (whenever I do stop for a cup of coffee at a National Trust tea room I inevitably spot someone reading that newspaper). While the National Trust itself attempts to reach out to inner-city youths, people from working-class and minority ethnic communities, the people coveted by the National Trust's marketing belong to an exclusively white England that is situated in the imagined past of the glorious British Empire.

Travellers

White, middle-class people love to talk about travelling as much as they travel. Most of the people I spoke with had been travellers at one time or another – going on extended vacations to a number of places, joined together by epic bus rides, train journeys or ferry trips. For most of the people who had travelled, the long (six months seems to be the average) stay abroad was a gap year event, something done when single, with close friends from school or from university. They saw it as a rite of passage, something fulfilling, and also something that was good fun. Some continued to have extended vacations in the traveller style, using all their leave from work to go off to explore South America or China. All of them continued to reference the travelling adventures as something that made them more aware of the planet and globalization, and the diversity of the world's local communities. None of them considered the inequity of the relationships and experiences – how easy it is for rich, white Westerners to travel across borders, to stay immersed in a friendly Westernized subcultural space, and to come back home with a few native artefacts and some amusing anecdotes, just like European explorers and anthropologists form the colonial age. It is impossible for the local people they meet to do the same thing as part of their leisure lives – they will be lucky to have any days off work at all, never mind a vacation to somewhere different. I know travelling isn't just something elite white Westerners do – there are minority ethnic Westerners who follow the same backpacking trails. But these folk are able to navigate their way through the complexities of the encounters because they are rich Westerners, part of the bourgeois classes, following in the footsteps of the white people.

In their edited collection *Touring Cultures*, Chris Rojek and John Urry (1997) present a series of research papers that demonstrate the crucial role of identity formation in travel and tourism. Making the distinction between being a traveller and a tourist is the most obvious act of identity making they identify – to be a traveller is to be at one with the exotic, familiar with the foreign, comfortable with the new; to be a tourist is to be a happy and willing victim of the package holiday industry, visiting hotels that do food you eat at home, staying in resorts with bars that sell your favourite drink, with shops that can be found in any mall in any town in any Western country. With the shrinking world of the last few decades, travellers go further to find new experiences, become restless at home until they have ticked-off the next continent. Travellers surf the net to read stories written by other travellers about the most authentic experiences, the real lives of the exotic foreigners, the restaurants where the locals eat the real local delicacies. The tourist, on the other hand, trusts in the guidebook and the company rep in the hotel to take them to sample something exotic. Both traveller and tourist are complex, socially constructed identities, yet both assume the identity of the locals is fixed, something bound to the place of the holiday, unchanged by time – this is assumed of non-white strangers and also of poor whites in countries where rich whites go to find cheap places to stay (such as the Czech Republic). This, of course, is untrue: locals are as complex and changing in their identities as the visitors themselves. The whiteness of the global traveller is constructed in the colonial encounters that inevitably occur in the experience. Middle-class, mainly white Westerners travel around the world and use their economic and cultural power to create local spaces that allow them to have a taste of the exotic in the comforting surroundings of other Westerners. So we can see white Australians and New Zealanders on the Inca Trail, hanging out with each other, drinking Coca-Cola and checking out Facebook at internet cafes; or in Bali, hanging out with each other, drinking Coca-Cola and checking out Facebook at internet cafes; or in Kathmandu, hanging out with each other, drinking Coca-Cola and checking out Facebook at internet cafes.

Tourist studies researchers (for example, Holmes, 2001) have shown how the traveller has become the ubiquitous icon of Westernization and globalization, the rich young adult funded by their parents, getting a tribal tattoo when hanging out in Australia (if they are travelling from Europe) or in London (if they are travelling from Australia). The travellers claim to be searching for authentic places and spaces, real locals and cultures, but what they get is what is provided to them by the tour operators, guidebooks and websites they consult. In other words, they go to the places other white Western travellers have been, they pay money to white Westerners like themselves who run trips to local sights, they eat strange meals so they can tell folks back home about it, they speak only English and rely on English-speaking native guides, and they swap stories about the strange and untrustworthy locals (Azarya, 2004). What these (mainly) white travellers assume is that their travelling gives them the ability to empathize with and understand local diversity. They say the 'normal' (white) tourist never gets off the beaten track, only ever sees a crude stereotype of local cultures if at all, and might as well be holidaying in a hotel or apartment back in the States, or England, or Australia (Aitchison, 2001; Azarya, 2004; Edensor, 2002). This is of course true – but the traveller fails to see that their own encounters are limited, exploitative and culturally essentialist. In all the countries visited by Western travellers, it is the white Westerner who is finding these lost, isolated, non-white natives – it is the rhetoric of the Western explorer all over again, going out far away to see the exotic non-Westerners who still live in some pristine state of nature. The locals do not live in a far-away place. They live at home – but the traveller forces their home life to become a backdrop to their travelling story, something colourful to tell the folks back in the West. The traveller forces the local to be fixed in time, to be untouched by Westernization or modernity, or hybridity. Where the travel tourism industry takes over elements of the traveller trail, this is done on Western terms, controlled by Western-ers for the benefit of other Westerners: the hedonistic lifestyle of white, Western travellers in Goa, for example, is a source of political tension locally (Saldanha, 2005).

So far travellers and travelling may seem profoundly monolithic, but there are in fact different approaches to being a traveller, which equate to different subcultural choices based on the politics and ethics of impact. For some travellers, the stereotypical white Westerner teenager with one eye on Facebook and the other eye on whichever person with whom they want to make out is something that they distance themselves from. Some travellers distinguish their 'serious' traveller sub-culture from the hedonistic, materialistic traveller subculture of the early twenty-first century. They accept that the majority of the young peo-ple who fly out from the West to the rest of the world for a gap year or an extended trek across the traveller trails are bad examples of late modernity and the instrumentality of tourism (and the hegemony of the West). These alternative travellers favour a Habermasian communicative rationality of counter-culture and interaction, an awareness of the pol-itics and an engagement with understanding the cultures they visit through sustained dialogue. Tourism and outdoor adventure researchers have used the idea of serious leisure to explore the tensions between long-term travellers or guides and tourists passing through or work-ing a season. Scott Cohen's (2010) research explores the way in which long-term travellers, those he terms lifestyle travellers – people who repeatedly travel for months at a time following the well-used routes away from the tourist resorts in India and Thailand – create a sense of identity for themselves through that travel. These travellers are of course Westerners. For some of his respondents, travelling in the Far East and elsewhere is something they earn through paid work in their countries of origin, to which they are forced to return to find jobs to keep them in the travelling lifestyle. For others, travelling allows them to find new work experiences in different cultures, or at least allows them to work in casual jobs associated with the travelling subculture (guides, bar staff, hostel keepers and so on). For most, the long-term travelling lifestyle is itself a job, at which they expend their energy and effort without much regard for any long-term plans of settling somewhere in a normal occupation. Cohen is sceptical of the ability of some lifestyle travellers to maintain a strong sense of self: many of his respondents in fact struggle to make sense of their place, despite their avowed love of travelling and their alternative lifestyles. This problem seems to stem from the fact of the (post)colonial encounter: no matter how much the white Westerners try to make themselves a part of the cultures and communities through which they travel, they will always have the specificity of their white, Western gaze and the inability to accept the complexity of the Other.

The tourist industry makes money from taking people to places where they can relax and forget about the stresses of their everyday lives. But some in the tourist industry claim tourism offers more than that. Some travelling holidays are explicitly sold to Westerners as opportunities to expand their horizons, to interact with some foreign culture and to become global citizens aware of their environmental impact. Tourists become sustainable travellers, limiting their carbon footprints to reduce the impact of global warming, helping out in conservation research in the polar-regions (helping scientist count polar bears and whales, for example) and learning about the place of the locals they meet. For these tourists, travelling becomes something more than just escape and adventure – the function of travelling is a way of becoming, a way of learning, a way of giving and taking. Scott and Becken (2010) edited a special issue of a journal that highlighted the tensions and paradoxes in sustainable tourism, from the (still) huge impact on climate change (despite the claims to the contrary) to the exaggerated claims about the function of sustainable tourism made by tour operators and those who pay for those tours. Sometimes some people do learn some things, but not often. More usually, sustainable tourism is another myth for rich, white Westerners looking to salve their consciences – hoping to make up for years of instrumental consumption and appropriation of resources (the year after year of vacations using environmentally disastrous jet planes) through an act of charity or mercy. Again, it is white Westerners

who are engaged in this sustainable tourism, who pay money to pursue more planet-friendly adventures. To mix a metaphor or two, they
may be reducing their carbon footprints but they are not reducing their
white feet, which are trampling over the local diversity and hybridity.
The sustainable tourism traveller insists on seeing native cultures that
have some fixed essence, because those are the things the Westerner
thinks are more authentic. So the sustainable traveller travels to Brazil
to visit tribes in the Amazon rainforest, and the tour companies make
sure the traveller's gaze falls on people from tribes who look and act
like the people from tribes Westerners expect: the Westerners want to
hear the locals talking about their hard life making fishing rods and the
beauty of their hand-carved drums, and are disappointed if the people
in the rainforest buy their fishing rods online and say their favourite
singer is Katy Perry.

Travelling, then, is a pernicious form of tourism, because it is sold
to Westerners who have the money to travel as a more ethical lifestyle
choice. They believe that they are doing something that challenges
Western imperialism, the instrumentality of the modern tourist industry
and the homogeneity of globalization. But the flows of travel and power
flow in one way only – from the white West, the home places of the
travellers, to the non-white locations they visit on their travels. There
is little flow in the other direction: rich elites from China and India
have started to visit the West as tourists, but this flow is not available
for the vast majority of non-Western, non-white people. Non-white,
non-Western, non-elite people who do try to travel to the West face
problems with visas and crossing borders, and are labelled as dangerous,
unwanted economic migrants and potential asylum-seekers. The reality
of the traveller experience is defined by the cultural capital and logic of
Westernized, global popular culture. Sophisticated, Western, white elites
can challenge their own power and their own supremacy through recog-
nizing the commodification of tourism and its unsustainable practices.
In doing so, they insist on shaping the Other as something permanent,
anti-modern, local and parochial, something that does not make any
significant contribution to modern, global culture or society. The Other
becomes something displayed in the pictures and videos posted to social
media sites by these Western neo-imperialists – the complex and fluid
identities of those patronized by the Western traveller are permanently
inscribed by the artificial gaze of the camera, forever the object of a
white, voyeuristic desire for atemporal diversity. The tourist does not
care for the exotic and foreign people and places she visits, but the
traveller claims to care deeply. Both are agents of white hegemony and

Western colonial interests, demanding that cultures change to fit what they think those cultures should look like, demanding that places are cleaned up so they are somehow more pure (more reflective of some essential mythology of the alien).

Conclusions

What should be clear from this critical analysis is that modern tourism, the product of Western power and global capitalism, is a dangerous leisure activity because of the unequal power relationships behind every encounter. Travelling is one part of this dangerous modern tourist industry that tries to make room for a Habermasian communicative rationality behind the choices and encounters made: the traveller insists she is not just a tourist, that she approaches local cultures with respect, that she travels abroad in a sustainable manner and that she minimizes her reliance on instrumental systems. However, like package holidays and other forms of foreign tourism, the entire experience is predicated on a belief that strangers are strange, and predicated on a white Western control of global travel flows. The only truly ethical tourism is that which does not involve crossing borders and gazing at Others. In a world where power is still in the hands of the white West, foreign travel for holidays – no matter the nature of that travel – is fundamentally racist in its assumptions about who gets to travel, who gets to look at strangers and who has power over their identity.

11
Whiteness and Outdoor Leisure

Introduction

In all the countries of the West, but especially in the United States and in the United Kingdom, horse-riding is associated with elites dabbling in mythologicized country living. In the United States, the cowboy is the archetype of the white frontiersman, fighting the Native Americans (the 'red Indians'), breaking the law (fighting the system) and smoking his cigarette (the symbol of his rugged individuality and disdain of soft folk back on the East Coast) (Fishwick, 1952). However, the myth of the cowboy has been transformed by the industrialization of ranching, the decline of the white rural working classes and the rise of equestrian vacations and clubs where elite whites can perform both the myth of the cowboy and the older elitism of colonial equestrian events associated with the old country (Ellis, 2011). Kate Dashper's (2010, 2012) ongoing research project exploring issues of equality and power in equestrian sports shows how elitist equestrian sports and equestrian leisure activities in the United Kingdom can be. Dashper (2012, p. 6) notes that in the United Kingdom:

> Women dominate equestrian sport numerically, with two thirds of all British riders being female (British Equestrian Trade Association, 2006). Popular associations of horses and riding also tend to construct it as a predominantly female activity within the UK. However, men compete successfully in all forms of competitive riding and tend to dominate the elite levels of the sport, especially in racing and show jumping (Cassidy, 2002). Equally, although current associations with girls and women may dominate popular perceptions of horses and riding, Birke and Brandt (2009: 195) discuss how equestrian discourse

is based on 'a predominantly masculine knowledge', and equestrian sport developed from military riding and was thus infused with masculine meaning. As a result of historic links with the military, the landed gentry and upper-class society, equestrian sport still has strong class undertones (Riedi, 2006). Although participation in equestrianism is now much more evenly spread across socio-economic groups and now involves more women than men (British Equestrian Trade Association, 2006), the history of the sport remains firmly linked to upper-class masculinity, and these associations continue to affect how the sport is perceived today.

In this particular paper, Dashper is interested in masculinity and sexuality and the ways in which gender is constructed in equestrian sports such as dressage. However, it is clear from the description and analysis of those who take part in horse-riding that such outdoor leisure is a site where hegemonic whiteness is at work. Those who wish to associate themselves with activities that legitimate and normalize the British elite classes – who hold on power in the British countryside – are attaching themselves to a power relationship made rich on slavery and oppression, which makes blood and family and inheritance and belonging and nationalism and race all ways of justifying the white control of the British countryside. Dashper is careful to point out that the images of whiteness and elite dominance in British equestrianism do not reflect the reality of a more equitable demographic of equestrian participants. There are white, working-class horse riders at all levels of equestrianism, for example – but the cost associated with owning and keeping a horse, the traditions and cultural capital needed to access such outdoor leisure and the connections between horse-riding and elite activities such as hunting and polo leave equestrianism a very white space and leisure form.

This chapter will focus on whiteness and outdoor leisure: leisure activities such as walking, climbing, cycling, canoeing, sailing, skiing, cross-country running and involvement in conservation volunteering. Outdoor leisure is a problematic term used to try to encompass the plethora of different leisure activities that take place outside of urban areas (though that does not stop some runners and walkers negotiating their way through cities, and it does not stop public parks in cities being sites of informal outdoor leisure activities such as strolling, having picnics and feeding the ducks). I use it because it is more inclusive than outdoor recreation – it allows me to include volunteering and being in the outdoors, and not just doing sport-like activities in the outdoors.

Of course, outdoor leisure, as leisure scholars use the term, excludes dark leisure activities – outdoor leisure is not to be mistaken for random sexual encounters in a park, for example. The term is used to describe specific physical activities and engagements with the outdoors – activities that are viewed as morally or spiritually good, healthy for the body and for the mind. These activities are associated with the rural, the countryside, what might be described as the wilderness: they are activities that involve some communing with nature. There is a blurring of definitional boundaries at the more formal, serious end of outdoor leisure, where such activities are best described as competitive sports. The first section of the chapter will survey the research literature on outdoor leisure to identify the presence or absence of whiteness in the theoretical frameworks used to understand outdoor leisure. The rest of the chapter will focus on long-distance walking in North America and the United Kingdom, and will use new primary research to show how such outdoor leisure practices are connected to the imagined, nationalist nature of instrumental whiteness.

I've done long-distance walks, done a few big hills and walk or run dozens of miles every week: walking I do with my wife or occasionally with friends; running I do alone, just me and my music and the countryside. All the white people I speak to have some outdoor leisure pursuit they do, either alone or with family or friends. They all want to leave the urban streets in which they live and work (or they have already moved from the city into the suburbs or into the countryside). When I ask some of the white people why they like these outdoor leisure activities they all tell me it's the feeling of being outdoors in the countryside, being healthy, being away from the miserable environment of urban Yorkshire. One cycles along the canal with his son every weekend, to get out into the fields around the edge of the city. Another lives close to the moor and goes out for a long walk every day – the benefit of being retired. These people I know are all in my extended network of contacts. They are mainly middle-class or upper working-class white people (people who have bought their own houses and lived the Thatcherite dream). Outdoor leisure seems to be fashionable, especially for people with young children. The families go out together, or it might just be the dad, and they camp, cycle, walk, ski. Living like I do in small town on the edge of the hills I see lots of people come through at the weekends and in the school holidays: teams of cyclists from Bradford or Wakefield or Leeds dressed in team Sky colours, eating bacon butties from the van on the High Street; walkers, proper ones in worn-out kit and the showy ones with the expensive gear who drive up and stroll through the woods; boaty people on the canal; and fell-runners who have come for a few days of self-imposed torture. I go walking myself along many of the footpaths around here.

Nearly everyone I see is white – I do often see some British Asian families going to picnic at Bolton Abbey (a popular beauty spot), but this is so unusual for the Dales to be noteworthy.

Research on outdoor leisure

Any examination of the research literature on leisure studies will show that outdoor leisure activities are the subject of a significant proportion of the scholarly activity. North American leisure studies have evolved from (and is still strongly connected to) two paradigmatic frameworks: recreation management, which is concerned with understanding how to manage outdoor recreation activities, and how to ensure people get access to such recreation opportunities; and what might be called the social psychology of physical activity, which is concerned with accounting for the motivations and choices of those who participate in serious, outdoor leisure (Coalter, 1997; Henderson, Presley and Bialeschki, 2004). Because of those traditions, it is easy to see why North American journals such as *Leisure Sciences* might be filled with papers on these outdoor leisure activities. But the interest in outdoor leisure runs across leisure studies, into Europe and Australia, New Zealand and the emerging leisure studies regions of South America and East Asia. This is partly due to the spread of North American ideas and research agendas (and the specialization of social studies of sport into its own subject field), but it is also due to the cultural turn in leisure studies, the shift from classical sociology to exploring subcultures and identities (Bramham, 2006). At the time of writing, the latest issue of *Leisure Studies* (31[1]) had two papers out of seven devoted to research on outdoor leisure. One of these is by two Taiwanese scholars, who explore surfing using the serious leisure theory of Bob Stebbins (Cheng and Tsaur, 2012). The other is an exploration of what the author calls the 'physical youth culture' of snowboarding (Thorpe, 2012).

The popularity of research on outdoor leisure also reflects the interests of the researchers themselves. Leisure studies, especially qualitative research associated with the cultural turn just mentioned, are dominated by research about specific cultures and communities of leisure to which the researcher belongs. There are plenty of leisure scholars who write dispassionately about leisure activities that they do not personally do or like – to engage with the spectrum of leisure activities in late modernity it is impossible to avoid writing about leisure activities you do not personally like. But leisure scholars are fortunate that usually they can write about the things they like, the things they are passionate about, the leisure activities they did before they became academics and the leisure pursuits they still do on a weekend. From a methodological perspective, this is one of the strengths of qualitative, cultural leisure studies: I could not do the research Dashper has done because I have no history of horse-riding, no involvement in it, no way of fitting in and understanding it beyond the cursory discussion of what the culture looks like from outside. For most leisure studies scholars, the leisure activities they choose to do research on are obsessions, or hobbies, or passions, or things they have become familiar with through family and friends. This obsessionality of leisure studies, while a methodological strength, is, for a number of reasons, an epistemological weakness. Firstly, there are multitudes of new, in-depth, ethnographic accounts of cultures, communities and scenes that merely add to the pile of similar in-depth accounts of slightly different groups – leisure studies are in danger of being overwhelmed by detail and voices and writers waving flags for their favourite leisure activities. Secondly, there is the problem that leisure studies become too specialized – football researchers, for example, publish in their own journals, attend their own conferences and can quite easily never come across any research that is about another form of leisure or another form of global industry. Finally, there is the problem that leisure studies merely reproduce the leisure interests of the white, Western middle-class leisure scholars who get careers and move up the publishing ladder – feminists have in the past rightly critiqued leisure studies for the dominance of middle-class men writing about the things middle-class men like; a similar critique can be levelled at leisure studies for the dominance of white, middle-class Westerners writing about their favourite leisure activities. Outdoor leisure, then, is a significant topic of interest in leisure studies because it is a white, Western middle-class leisure form. It is the kind of leisure that involves significant economic and cultural capital to do, and the kind of leisure that is associated in the West with a retreat from the multicultural spaces of the cities (Butler, 2004).

In research on outdoor leisure activities there is evidence of a range of identity formations. One of the key identities associated with the outdoor leisure scene is the serious adventurer, challenging one's body to achieve goals. This identity is often gendered: typically, it is the male rock climbers who try to outdo each other and conquer the rock (Heywood, 2006). This hegemonic masculinity (see Chapter 10, where this term is defined) is tempered by the respect for nature and wild spaces that is attached to the culture of climbing (and outdoor leisure more generally). There are environmental concerns about some outdoor leisure, such as the impact of skiing resorts or the rubbish at Everest Base Camp (now that climbing the peak is another corporate tourist adventure). However, rock-climbers treat the spaces in which they climb with more respect. Novice climbers learn from older climbers – and from climbing books and magazines – that the activity has always been associated with finding peace in the wild, a spiritual activity that supposedly allows individuals to be in tune with their surroundings. Despite that green identity, there are still debates about damage on rock faces and safety kit pinned to the rock, as different climbers perform different identities: the traditionalists who spurn fancy kit for pure rock climbing; the route-baggers who want to get up the rock safely; and the manly conquerors who use any means to secure success and accolade.

The climber is also hegemonically white. The history of modern climbing started with white elites in the West in the nineteenth century seeking out and claiming the first ascents of rock faces in mountain areas – even where Aboriginal peoples might have already lived in those areas and climbed the rock faces themselves. The white, elite Westerners had the capital and the leisure time to invest in climbing expeditions – as well as the benefit of technologies that allowed kits to become lighter and safer. Climbing became a white, working-class pastime in Europe and the United States by the first half of the twentieth century, and there are traditions of working-class climbers going out during weekends to places such as the Peak District of England. But climbing continues to be a white outdoor leisure activity: white folk still go climbing to escape urban places and find like-minded white people. And the white elites of the West still travel to exotic locations to 'bag' mountains and routes, planning their trips like colonial adventurers and paying locals to carry their luggage and guide them to the places where they can start the climbs. Going out to the strange Other country and returning with tales of new mountains (unclimbed by any white people) are still part of the narrative of modern-day climbing. In contrast with the white rurality of climbing is the metropolitan, urban physical activity of parkour, of free running (Saville, 2008). Parkour is the activity of running, climbing and leaping across urban spaces, using the skills and strength of gymnastic movements to provide an aesthetic of style which is fluid, pleasing to the eye and seemingly challenging to the laws of gravity. The sport emerged in the large estates on the fringes of French cities such as Paris and Marseilles, where large numbers of white and non-white poor people live marginalized from mainstream French society. Parkour was associated with graffiti artists, skateboarders and urban cyclists, young people (mainly men) living in a subculture inspired by urban America

and the North African diaspora. It has become an everyday activity performed by young people in town and city centres across the world.
At the same time, it has developed formal rules and competitions – but
this development has not altered the sport's perceived coolness or rebel-
lious nature. Parkour performers and professional competitors alike take
risks without safety mats, ignore local bye-laws on trespass and use of
spaces and see their activity as something that subverts architecture and
plays with notions of access and consent. What this demonstrates is
the way in which similar activities become different types of leisure –
outdoor leisure is white when it is connected with an escape from the
city and a search for a traditional countryside.

Skiing is one of the most obvious exclusionary outdoor leisure pursuits. Participants need to have time and money to train to be com-
petent. They need to be able to purchase expensive equipment and
holidays. Skiing developed from a rural practice in late modernity into
a leisure activity associated with the aristocracies of Europe and the
West. Ski resorts were opened throughout the West as places of retreat
for the elite white classes, where they could demonstrate their hegemonic power through staying in exclusive villages and hotels for weeks
at a time. The bourgeoisification of the West saw skiing become a popular marker of distinction for the middle classes, especially in countries
such as the United Kingdom. Although the Westernization of the elites
in many countries has spread the popularity of skiing throughout the
globe, it remains a mainly white, bourgeois and elite leisure form.
In their research on the tensions between skiers and snowboarders,
Edensor and Richards (2007, p. 103) characterize skiing's elite nature
in Western modernity in terms of cultural capital, class and *habitus*.
They argue:

> Skiing has long been characterised by an elitist, exclusionary identity.
> The cost of lessons, clothing, equipment hire and skiing holidays has
> limited participation to an affluent elite, for whom it has become
> a fixture in the social calendar, part of a broader strategy that dis-
> plays status through an association with particular practices and
> places. To confirm that skiing is the preserve of the select few and
> emphasise their disapproval of this exclusivity, snowboarders fre-
> quently cite the Royal family's patronage of elite skiing venues.
> As a form of 'conspicuous leisure' (Corrigan, 1998: p. 17), skiing
> produces 'gains in distinction' (Bourdieu, 1991: p. 362), and has
> physically and through 'techniques of sociability' (Heino, 2000:
> p. 177) excluded snowboarders as inappropriate members of the

snow resort community. Whilst these strategies are mobilised to maintain a class-specific 'cultural capital' within the field of (winter) sports, snowboarding mobilises its own forms of distinction, better typified as 'subcultural capital', wherein distinction is achieved and maintained through displaying performative skill, authenticity and 'cool'.

(Thornton, 1995)

Unlike skiing, snowboarding is seen by its participants as belonging to a more equitable, postmodern world of leisure lifestyles and leisure choices. As Edensor and Richards (2007, p. 100) state:

Although Heino refers to the 'still contestable history of snowboarding' (2000: p. 177), the pursuit is most evidently an adaptation of skateboarding culture, which in turn, evolved out of surfing culture. It is, according to Kevin Taylor, the director of the Tamworth Snow Dome, 'an easy conversion from skateboarding to snowboarding...a natural progression, because the equipment and manoeuvres were already familiar to kids on the street'. However, the rather multi-ethnic, multiclass, primarily urban skateboarding culture has not been duplicated within snowboarding, which is overwhelmingly white and middle class, and primarily pursued by males, although according to Heino (2000), less so than other lifestyle sports. Yet like many lifestyle sports, snowboarding is also a hybrid culture which combines aspects of grunge, punk, and hippy cultures and more latterly hip-hop and dance, amongst others. In this sense, it is exemplary of the intensified fragmentation of contemporary youth cultures and their tendency to adopt numerous styles and bring them together through bricolage.

(Bennett & Kahn-Harris, 2004)

Skateboarding belongs to the urban world of diversity and multiculturalism, but snowboarding's whiteness is due to the rural and commercial nature of the particular outdoor leisure form. Like skiing, snowboarding is a commodified activity that takes in holiday locations (where there is snow, where there are slopes) or on artificial ski slopes (which are expensive to access). There are very few snowboarders who can hang out on the street and engage in their chosen outdoor leisure activity. So the cultural capital of snowboarding remains essentially similar to that of skiing, and hence the *habitus* of snowboarding – despite its origins in youth, urban, counter-hegemonic subcultures – is white.

One has to have money, and time, and access, and understanding of the rules of ski tourism: all things that mean snowboarding is likely to be whiter and more bourgeois (albeit hipper and younger than skiing) than skateboarding.

Similar dynamics of hegemonic whiteness and identity formation can be found in other 'lifestyle sports', outdoor leisure forms such as surfing, windsurfing and canoeing (Wheaton and Beal, 2003). These activities are viewed as being hip, countercultural activities, but they remain fixed in a metropolitan, bourgeois, white habitus. They are expensive hobbies, and they give the younger, urban, white middle-classes a place to get away from it all by finding peace and satisfaction in the water. In this regard, such 'cool' water leisure activities resemble the more established (and more white, more elitist) water sports such as sailing and boating. In surveys of active participation in sports and physical activity, swimming often appears at the top of the lists as the most popular. This is often used by national governing bodies of sport (such as the Amateur Swimming Association in the United Kingdom, or USA Swimming in the United States) to make the argument that they are being successful in developing their sports and need to be rewarded with more money for their development and performance schemes. Swimming as an elite sport demands hours of training every day from its participants – and this means only those young people with parents able to spend time and money taking them back and forth to the swimming pools become successful (so the whiteness of elite swimming is explained by the sociological advantages of class and has nothing to do with spurious scientific notions of black physicality). In fact, most people who swim do not swim under the auspices of national governing bodies for swimming: very few swimmers are members of swimming clubs or involved in competitive swimming. Most people who swim do it for fun, to be healthy, or to be with their families. They swim in public sector swimming pools, which are often designed with special effects such as wave machines with children (and their dollar-paying parents) in mind. Or they swim in private gyms to keep fit; or they swim at school; or on vacation at an adventure park or hotel; or in the sea. Some people even choose to swim in the wilderness, finding satisfaction in swimming rivers, lakes and streams (Deakin, 2000). This counter-movement tries to reclaim the experience of swimming in water from the rational recreationists and the corporations who have turned swimming into paying. Swimming, then, is emblematic of liquid modernity, with a fluidity of meanings given to it, with most swimming experiences taking place away from the institutions and rules of national governing bodies. Such countercultural activity, however, is available only to those who have the cultural capital and confidence of the white, metropolitan middle classes.

Walking

All tourism has an unwanted commercial and environmental impact on the destination, and the journey to the destination is often damaging as well. One way of making vacations less instrumental (and hence less damaging), and more communicative, is to walk out from your home to some nearby wild or rural space. Multi-stage, linear walks allow people to leave cities and enjoy the wild spaces on their doorsteps. It might be that a bus or train ride is necessary to get you to the suburbs, from where you can start your wild walk – that is acceptable, though obviously not as good as the feeling you get walking from your house out to some wild place. Most wild places have places to stay – in Europe, there are villages with hostels and places to eat and drink; in truly wild places there are often huts in which you can bed for the night. Communicative walking involves minimizing one's impact on the wilderness, spending money on local businesses and enjoying the feeling of being away from the instrumentality of the urban landscape. In the wild, you can choose where to walk, you can explore new areas and you can feel free. In the wild, you can walk alone, communing with your own thoughts,or you can walk with trusted companions, close friends and your loved ones. Either way, you can contemplate the wild spaces and realize the madness of the modern, commodified life that we all live, and for a day or two you can forget the tide of commerce and bureaucracy that threatens to drown us all.

The outdoor leisure activity of walking – sometimes called rambling or hiking – is a product of the industrialization and urbanization of the West. The rich, white elites of the West owned land in the coun-tryside and engaged in the traditional feudal sports of hunting, fishing and shooting. Walking became first a bourgeois outdoor leisure pursuit with the advent of railways and the rise of the white, Western, urban middle class. These middle-class white city-dwellers, captivated by the nineteenth-century trends of romanticism and nationalism, wanted to visit the countryside and find some rural idyll that connected them to some numinous past. They wanted to be in the countryside and of the country, to be like their rulers, but they could not afford to live in the countryside. So instead they compromised their need to live in the city, close to work, and their desire to be in the rural outdoors, by exploring public footpaths and other byways around small settlements and railway stations – in the Appalachians, in the Blue Mountains, in the Alps, in the Lake District. Walking in the countryside became a popular middle-class habit, and soon a tourist industry established itself in the wild and rural areas of the West, catering to the demands of walkers (Cormack, 1998). With the growth of the working classes and the availability of days free from working responsibilities, walking became a mass participatory activity (along with cycling). Walkers from cities across the West went out in their thousands on circular walks in areas of natural beauty. As the pastime grew in popularity, and the white working classes of Europe achieved the opportunity to take a week away from work, people camped or hostelled so that they could walk longer distances during their vacations. In the first half of the twentieth century, measures were taken by planners and policy-makers (throughout the West, but especially in countries such as the United States, where much of the land was rural, and also in mountainous countries such as Switzerland) to open up wilderness and rural areas to walkers through the establishment of sign-posted routes or the creation of long-distance walks. If the white, working-class walkers of the West were limited by time and resource to shorter walks in their own countries (often including a camp in the wild), the middle classes and the new white elites of the West were able to pay for long-distance walking trips in far-off places, often with guides and accommodation in comfortable walking huts.

Walking has declined in popularity among the white working classes in the West, but it remains popular among the white middle classes, who now use walking as a way of demonstrating their status and distinction. In the United States, this is done through going on long vacations along the Appalachian Trail or one of the other long-distance trails that demand knowledge, know-how, confidence and money (Littlefield and Siudzinski, 2012). In the United Kingdom, walking's status as a white middle-class leisure activity is established early in people's lives, through its connection to the Scouts and schemes such as the Duke of Edinburgh awards (Allison, Davis-Berman and Berman, 2012). Both the Scouts and the Duke of Edinburgh awards scheme reach out to minority ethnic groups, but the fact that they are reaching out to those groups demonstrates the whiteness (and middle-classness) of these two outdoor leisure networks. Walking is equated with moral superiority, clean living, healthy bodies and a respect for the history and natural environment of what is called 'The Great Outdoors'. Walkers in adulthood learn about the importance of the National Trust in preserving landscapes for the nation and can learn about walks through reading that organization's publications (see Chapter 10 for the National Trust and heritage tourism).

Most walkers learn about the places they can go to walk through initially talking to others, but they soon realize that the best way to find out about walks is through reading magazines and guidebooks (and web-sites). There are two types of walking magazine in the United Kingdom: serious and sedate. The first type is for serious walkers looking to stretch themselves across peaks and long stretches of moorland, in which young, white reporters talk about driving up to Scotland at 4am to bag Munroes (high hills), and young, white models in ads dress in gear that costs thousands of pounds. The second type of magazine is for older white people, or less fit white people, or white people with dogs or young children, who want to ramble around the classic British countryside and have a nice tea. Both types of magazine serve to exclude the working classes and minority ethnic groups of the country from feeling they would be welcomed at the top of a Munro or in the silence of a tea room.

Walking guidebooks in the United Kingdom are all influenced by the work of one person: Alfred Wainwright. Anyone who writes a guidebook writes in the shadow of Wainwright; anyone who reads a guidebook will almost certainly compare the quality of the information and the mapping to that provided by Wainwright (Palmer and Brady, 2007). From an audience of white, working-class fanatical walkers to an audience of white bourgeoisie finding in him an archetypal old English country man (the plain-speaking, awkward but fair chap who loves nature and uses the purpose of writing his books to escape 'her indoors'), Wainwright has become the old white English man of British walking. Although his own prejudices rarely appear in his guidebooks, his writing and his pictures represent an England and Englishness that is naturally at home in the hills, suspicious of modernity and respectful of traditions. Wainwright has given white, middle-class English people a way of finding their white Englishness in the silence of the English Lake District (an irony, of course, considering how the canonization of Wainwright and his books means every road and village in the lake District is full to the brim with middle-class white people reading his books). At the end of *Book Seven: The Western Fells*, the final pictorial guide to the Lakeland fells, Alfred Wainwright ends the series with 'some personal notes in conclusion', in which he writes (Wainwright, 1966, Some Personal Notes, p. 8) in a tone that brings to mind the most egregious verse of the Romance poets:

The fleeting hours of life of those who love the hills is quickly spent, but the hills are eternal. Always there will be the lonely ridge, the dancing beck, the silent forest; always there will be the exhilaration of the summits. These are for the seeking, and those who seek and find while there is yet time will be blessed both in mind and body.

Wainwright is not a great writer, or a sophisticated one. But his writing, and more specifically his pictorial guides, captured the sentiments of generations of walkers. Alfred Wainwright's life is well-known to those who go fell walking in the United Kingdom. The briefest facts are uncontrovertibly established by reference to the website of the Wainwright Society (http://www.wainwright.org.uk/about_aw.html, accessed 13 May 2010), where it is noted in an article written by John Burland:

> Alfred Wainwright was born on 17th January 1907, the youngest of four children at 331 Audley Range which was about a mile from the centre of Blackburn. His father was a stonemason who was unemployed for long periods and was also a drunk, for sometimes quite long periods. His mother was totally different. She was a god-fearing and hard-working woman and she brought him, his brother Frank and his two sisters, Annie and Alice up in decent, respectable if rather poverty-stricken conditions.

Wainwright's formative years are encapsulated by the archetypes of northernness in this biography. His feckless father and proud mother are described as if they have stepped out of the Lancashire soap opera Coronation Street. It is a hard, white, working-class upbringing, in a hard, white, northern mill-town. But Wainwright managed to transcend that life: through education, which offered him a way to become a professional clerk in the public sector. His conversion to fell walking is recounted as the article continues:

> His first encounter with the wonders of Lakeland happened on the 7th June 1930 when as a young man of 23; he travelled with his cousin Eric Beardsall from their home town of Blackburn, changing buses at Preston, to eventually arrive in Windermere. It was Whitsuntide and they ascended the lane from Windermere Bus Station to the nearby summit of Orrest Head, a hill 780 feet high overlooking the town. What he beheld from the summit was a scene of loveliness, a fascinating paradise, Lakeland's mountains and trees and water spread out below him. He described this experience as follows *'I was totally transfixed, unable to believe my eyes. I had never*

seen anything like this. I saw mountain ranges, one after another, the nearer starkly etched, those beyond fading into the blue distance. Rich woodlands, emerald pastures and the shimmering water of the lake below added to a pageant of loveliness, a glorious panorama that held me enthralled. I had seen landscapes of rural beauty pictured in the local art gallery, but here was no painted canvas; this was real. This was truth. God was in his heaven that day and I a humble worshipper'.

Again, there is a poor, working-class, northern life (changing buses at Preston), but this is contrasted with the aesthetics of the Lakeland fells ('the wonders of Lakeland', as Burland puts it). Wainwright's recollections are suitably evocative and rich with a nostalgia for a time when such a view was fresh: his recollection also suffers from a distinctly autodidactic, petit-bourgeois provincial style ('a pageant of loveliness') that was not fashionable in 1930, let alone 1966 or 2011. The reference to God in his heaven, of course, also reflects the strain of Christian piety associated with white, northern identity (Kirk, 2000). The article continues:

In 1941 he moved from Blackburn to Kendal to take up a position with the Borough Treasurer's department in the town, rising to the rank of Borough Treasurer in 1948. Virtually all his spare time was taken up walking the Lakeland fells until he knew them intimately.

Kendal is a town in Westmorland, on the edge of the Lakes. By leaving Blackburn for Kendal, Wainwright migrated from what would become a key town in the emergence of post-colonial, multicultural Britain to one that remained as white as its famous mint-cake. Wainwright worked most of his life in the public sector – even after he became famous. In 1952 he started working on the guides that would make him famous (discussed in the next section). At first, he was a reluctant celebrity: in the later pictorial guides, he even expresses some disdain for those who try to seek him out in the Lakes (Wainwright, 1966). By nature, he was a solitary man, with a reputation for terseness and impatience with those he did not like (Davies, 2002; Mitchell, 2009). However, towards the end of his life he became involved in making and starring in television programmes based on his walks. These helped transform the popular memory of his life and his work: after his death, he has become an icon of hill-walking and the Lake District in particular, with regular hagiographies and fawning adulation in television programmes (Robson, 2003).

Wainwright's fame is entirely due to the publication of the seven volumes of his *A Pictorial Guide to the Lakeland Fells* (Wainwright, 1955, 1957, 1958, 1960, 1962, 1964, 1966). These guides cover 214 high fells of the Lake District, and offer the reader a range of routes up and down every peak – sometimes following existing public footpaths, often following routes that have no legal status, and in some instances (such as The Nab in Wainwright, 1957) indicating that trespass was unavoidable. Wainwright's labour of love is a deliberate attempt to make readers become enchanted by the landscape of the Lakes. As he puts it in the Introduction to Book One (Wainwright, 1955, pp. 1–2):

> Surely there is no other place in this whole wonderful world quite like Lakeland... no other so exquisitely lovely... This book is one man's way of expressing his devotion to Lakeland's friendly hills. It was conceived, and is born, after many years of inarticulate worshipping at their shrines. It is, in very truth, a love letter.

Wainwright's guides, however, are more than tools for route-finding up mountains. Each page of the guides is drawn and written by hand. Each book is divided into sections based on individual fells: there are no page numbers to guide the reader; instead, one must find one's fell then find the relevant page in the section on that fell. Wainwright's sketches were unique at the time, though they have since become common-place in walking books: for every fell, the routes up to the top are marked on a number of elevations and a more standard map, with points of note written around the sketches. Wainwright provides sketches of the views from the tops of many of the fells, and many of the route descriptions are accompanied by long, idiosyncratic digressions on folklore, industrial history, ruminations on the nature of the landscape and acerbic comments about those visitors who do not respect the silence of Wainwright's beloved countryside (Wainwright, 1960, Personal Notes, p. 3):

> Great Langdale is a growing problem... Coaches, cars, caravans, motor-bikes and tents throng the valley... Some of the characters infesting the place at weekends have eyes only for mischief. These slovenly layabouts, of both sexes, cause endless damage and trouble.

In addition, Wainwright offered sketches of 'lovely' views of the mountains themselves, drawn in simple lines that accentuate the curves and ridges of the tops, and the lushness of the valley bottoms. The guides express a love of walking, and a distaste of those who spoil the fells – planners and unruly plebeians. When the first volume was published, it quickly sold out, and Wainwright's publishers were happy their investment was safe (Davies, 2002). Before he had finished the series, Wainwright's name had already become well-known among walkers, and many in the Lakes tried to identify him as he went about his business (Mitchell, 2009).

After he had completed the guides for the Lakeland fells, Wainwright turned to other places, including the Yorkshire Dales (Wainwright, 1970) and the Pennines. But it was his seven volumes on the Lakeland fells that remained his greatest achievement, and the one that earned him some immortality: the Wainwrights of the Lakes are as well-known as the Munros of the Scotland, and the pictorial guidebooks have never been out-of-print. Palmer and Brady (2007) argue that Wainwright's pictorial guides establish a particular landscape aesthetic: one that endows the rugged, rock-strewn high peaks of the Lakes with significant value, compared with other landscapes of the English countryside. Their analysis of the pictorial guides identifies Wainwright's attempt to normalize as an objective truth his own subjective opinion on what is beautiful. This objectification of his particular landscape aesthetic is then reproduced in the consumption and use of the guidebooks by generations of fell walkers. Wainwright's florid writing style, combined with the hyper-realism of his sketches, constructs an ideal Lakeland where the sun always shines, and the view from the tops is always wonderfully clear. The mundane realities of the English weather, and the long experiences of walkers getting lost on fells like Farfield in the mist, does not seem to diminish Wainwright's aesthetical objectification of Lakeland's landscape, or our desire to experience that idealized landscape though (re-)reading his books.

Wainwright's books were published at a key moment in English social history: when hill-walking and long-distance walking were still seen as part of a working-class social movement (Snape, 2004), but a time of increasing individualization and privatization of leisure practices (Borsay, 2005). Walking had become fashionable across classes, and a combination of legislation and motorway building made the Lakes more immediately accessible. Pressure to develop long-distance walking in the early twentieth century came from the same working-class lobby as those seeking the 'right to roam' – those seeking the freedom to walk in the hills and open spaces of the country, which were almost all closed to ramblers by wealthy landowners. The struggle between those who wanted to enjoy the freedom of the hills and the landowners who wanted to make profit from shooting rights was a key moment in English working-class history. The ramblers who fought the landowners over the right to roam also challenged those same landowners over access to hill-tops and wild-camping sites on long-distance walking holidays. After the Second World War, the establishment – in the United Kingdom – of National Parks led to the recognition of rural spaces as tourist spaces (Merriman, 2005). This recognition of course also constructed rural tourism as an alternative to agriculture as an income-generator in rural economies. Also, this recognition led to the idea of walking as a holiday, and walking as leisure. Long-distance walking holidays became a legitimate tourist pursuit, bringing walkers into National Parks and other rural areas: whether areas of some recognizable aesthetic quality desired by walkers, or areas in which local authorities tried to capitalize on the pursuit of long-distance walking by attracting walkers to revitalize economies marginalized by the effects of post-industrialism (Bold and Gillespie, 2009; den Breejen, 2007; Morrow, 2005).

My local countryside – the Yorkshire Dales and Pennines – has been transformed by the leisure and tourism industry associated with walking. Machin (2008) stresses the importance of the rural, semi-wild landscape of the Dales and Pennines in the history of Yorkshire tourism in the twentieth century as places to escape from industry: what Whyte (2007) calls a 'refuge' from the real world. Popular guidebooks to Yorkshire initially stressed the therapeutic nature of spa towns such as Harrogate and Ilkley (Machin, 2008). However, walking in the hills quickly became one of the primary reasons for visiting the Dales and the Pennines. Managing and developing tourism in the Dales was inextricably linked to the use of the landscape for leisure activities, walking in particular (Cochrane, 2008). Services developed, then, to encourage and support walkers, and local agreements with the Duke of Devonshire in the 1960s allowed walkers to gain access to Barden Fell and Barden Moor (above the day-tripper magnet of Bolton Abbey village). As Thomas (2008) suggests, walking is one of the key income-generators for tourism businesses in the Dales, but the footpaths in Yorkshire Dales and Pennines are nowhere near as busy as those in Lakeland. Walking guidebooks to the Dales are nowhere near as iconic as Wainwright's pictorial guides, do not feature in television programmes, and the names of their authors remain largely unknown outside the walking scene.

In the second half of the twentieth century, two long-distance walks that cut across the Dales and Pennines were devised. Both became relatively popular, and have helped create a number of tourist services along their routes such as hostels, bed-and-breakfasts, pubs and bag-carrying services. The Dales Way begins at Ilkley and ends at Bowness on Lake Windermere. On the way, the path goes up Wharfedale, past Bolton Abbey, Grassington, Kettlewell and Buckden, before climbing over Cam Fell and down into Dentdale. Passing Sedbergh the route follows the River Lune for some distance before crossing the water and leaving the old West Riding of Yorkshire to enter the agricultural land between the Dales and the Lakes. This route was developed by Colin Speakman of the West Riding Ramblers' Association in 1968, and although not a National Trail it is recognized and supported by the National Parks and Local Authorities through which it runs.

The second long-distance walk in the Dales and Pennines is the Pennine Way, the first of the country's official National Trails. Opened in 1965, it was inspired by Tom Stephenson's Pennine Trail (Stephenson, 1989). It runs from Edale in Derbyshire to Kirk Yetholm in the Scottish Borders. Its journey through Yorkshire begins on the watershed of the Pennines south-west of Huddersfield. It passes Marsden, Blackstone Edge, Heptonstall, Top Withins near Haworth, then Lothersdale, before coming to the old West Riding town of Earby. Then it continues north through Gargrave, Malham, over Pen-y-Ghent into Horton, and beyond to Hawes, Great Shunner Fell, Swaledale and Tan Hill Inn, then the edge of the old North Riding in Teesdale. The Pennine Way is a real challenge to walkers, who usually take 15–20 days covering its 268 miles: in the 1960s and 1970s, before many of the worst bogs were made safe through diversions and the laying of stone causeways, it was even more arduous. It is on this version of the Pennine Way that Alfred Wainwright published a guide in 1968 (Wainwright, 1968). Both these long-distance walking routes give their users the chance to escape from the urban, to escape from the tourist crowds, to find silent moors, valleys and hilltops. They allow their users to find themselves and find a slower way of life, a rural England, a wild, untamed Yorkshire. There is nothing wrong in wanting any of those things. But it is problematic that the people who feel the need to do all these things are mainly white, and almost certainly middle class. Some of the desire to walk in this landscape must be connected to a mythological past, nostalgia for a white world of rationing and Empire. Some of it must be associated with the continued sense of confidence and superiority drilled into elite white British children in school, the reality that they still hold economic and cultural power in a late modern, globalized Britain. And some of it must be associated with the white sense of privilege and belonging that desires to own the land and keep the land for those who have a psychogeographical relationship to it.

Conclusions

Outdoor leisure activities are dominated by white people – in the West, domestically; and in the wider world through the power of white elites taking adventure holidays in 'exotic' or 'untamed' places. It is clear that outdoor leisure practices such as walking are connected to the imagined, nationalist nature of instrumental whiteness in the West. By walking or cycling or climbing across national and regional landscapes, white people demonstrate their ownership of the national, racial, white myths of Western countries. This whiteness is instrumental – it is a product of the imperial and capitalist hegemonies of the twentieth century, and their continued hegemony across much of the world in this century. Outdoor leisure is connected through a series of translations with history, historicizing discourses, legitimate ownership and belonging, insiders and outsiders, the survival of the pre-modern, pristine nature, purity and authenticity, the importance of the land as a signifier for purity and the ersatz and hybrid nature of late modern cities, and the power of white elites. For less confident white, Western social groups, the rural and outdoor leisure might be indicative of some residual cultural power, a retreat from the reality of modern life and the multicultural street to the monoculturally white footpath through the fields. But that retreat is predicated on the instrumental power of white elites, which maintains the countryside and the rural as the well-spring of nation and the nation's people, a sacred glade protected from the influence of foreigners and the Other. Non-white people in the West obviously do visit the countryside, and they do many of the leisure activities mentioned in this chapter. But their activity and visits are controlled by the cultural capital they need to become middle class or elite, as well as the racism and prejudice they face and fear. The countryside and outdoor leisure activities remain white in the way they create instrumental, hegemonic whiteness, nationalism and myths of belonging – and in the way they continue to be activities enjoyed by white people when they want to leave the cities behind.

12
Conclusions

In this conclusion I provide a synthesis of the key findings of the rest of the book and return to the problems raised in the first third of it to show that the seemingly inevitable relationship between modern leisure and whiteness (and whiteness and intersections of class and gender) is not a necessary feature of modern life, only a contingent one. First, it is necessary to return to the question of whether the world has changed and become something postmodern, something where multiculturalism has triumphed and people readily accept difference, diversity and hybridity.

The world has changed since the advent of modernity. The British Empire has fallen. American economic power is challenged by countries that used to be dismissed as problem cultures. Global flows of people and ideas have changed the way people think and act. Women have been granted legal protections guaranteeing them equal rights in many countries, and these rights have been extended to minority ethnic groups across the West. Populations have changed with global migration flows in and out of Europe. Westerners think they live in countries where individuals can achieve anything they want without suffering any kind of discrimination that holds them back. Minority ethnic individuals in the West have achieved success in the workplace and in politics, resulting in the rise of a minority ethnic middle class in countries such as the United States. Minority ethnic individuals read the news on television, star in films, play soccer for their country and become globally renowned musicians or authors. One could be forgiven for thinking that modern society has changed so much that the old power relationships and social structures have vanished. Certainly, this is the view of some postmodernist scholars (for instance, McGuigan, 2006) who suggest that we have become postmodern precisely because the old hegemonies have been overturned. However, it is clear that the evidence that some things have changed and some people have more political and social agency – and some of the more brutal forms of discrimination have been outlawed – is only evidence that the world today is different from the world of the nineteenth century. The transformation of laws in the West, and the universal concern with human rights, is something to be recognized and applauded. But, as this book has shown, there is no level playing field, no equality of opportunity, because hegemonic power relationships define how likely anyone will succeed in life. The elites of the West have slowly and reluctantly conceded ground to the human rights movement and have loosened the tests and rituals that give individuals entry into their inner circles, but they still control who joins them and still act in the interests of their hegemony. In global politics, this is seen in the way governments and financial corporations work with transnational organizations such as the World Bank to keep poor countries in debt, or to turn poor countries into ruthless free-trade zones where Western companies can accrue extra profits while public services in those countries are sold-off or cut (Klein, 2000). In the United Kingdom, the hegemonic powers ensure that a small class of people educated in private schools and at Oxford and Cambridge dominates every aspect of senior management and leadership in society: the media, the Church of England, the City of London and the financial sector, the government, the old establishment associated with the Crown and the civil service. Some (a tiny fraction, but some nevertheless) of these people might be from minority ethnic backgrounds, they might be women, they might be from working-class families – but they have all learned how to protect the values and aims of the ruling elites, how to identify themselves with that elite and how to look out for other individuals from that elite who might be looking for a job.

Instrumental whiteness is shaped by modern leisure, and in turn instrumental whiteness shapes modern leisure: restricting choice while acting to create the illusion of choice, protecting hegemonic interests, racializing and marginalizing Others while acting to create the illusion that white hegemony is not being constructed and protected. In popular culture, the television and films we watch trick us into thinking that we live in an age of unrestricted opportunity, while at the same time they normalize the norms and values of the instrumental white hegemony. In music, taste is used to differentiate between classes, to provide a sense of distinction and authenticity for the elites and middle classes – which makes distinction act to support instrumental whiteness. Pop and rock music are racialized, with black music deemed by consumers as less authentic than white music, and black musicians restricted to stereotypes about brutality and criminality. In sport, the myths of racial science continue to be evoked to account for the success of black athletes in some sports, while sports themselves become racialized: basketball becomes a sport 'suited' for black people; cycling and horse-riding become sports for white people who have the resources to buy equipment and time, and who wish to demonstrate their ability to accrue hegemonically white cultural capital from a sports culture that was created by white elites for white elites. The sports-media complex serves to reinforce the racialization of sport while at the same time acting as a propagator of the myth that sport (and society) is 'colour blind', the level playing field of the myth of Western neo-liberalism. Imagined communities and nationalism operate to perpetuate notions of Western superiority and Western exclusivity, while sports commentators continue to racialize black athletes. In everyday leisure, the tacit rules and informal networks of cultural capital act to make the particular habits of the white, Western elites become universal, unquestioned and reinforced with every cup of coffee sold. Ethnic food is still food from minority ethnic cultures, food from exotic cultures outside the (white) mainstream – as if any food could be non-ethnic (all food, after all, has its culinary origins somewhere). Tourism takes white elites from the West into encounters with the exotic, with the Other, forcing locals to adapt and adopt forms of being suitably neutered for the edification of the white tourist. The tourist industry sells myths of authenticity, of heritage, of exclusivity, while hiding the mechanism of hegemonic, instrumental racialization. In outdoor leisure activities, even things such as walking and climbing are racialized: whitewashed and shaped to preserve and protect white interests, while acting to hide the act of whitewashing. Throughout this book, we can see what happens when leisure comes together with instrumental whiteness and other forms of hegemony such as masculinity and class. The notion of leisure as the activities done when someone has some free time to relax and pursue something that makes them feel happy is far removed from leisure in this book. Leisure here is itself a product of Habermasian instrumentality (Habermas, 1984, 1987): it is a servant of the white, Western, elite men who made modernity (Spracklen, 2009, 2011), and whose descendants still wield enormous power in this globalized period of later modernity.

Whiteness is a feature of every aspect of leisure in the twenty-first century, and it is an important process of identity-formation in the Westernized, globalized world in which we now live. Modern, Western society is constituted on a myth of individualism and meritocracy, but it is in fact dependent on continuing dynamics of racialization: the construction of whiteness and the construction of the Other. This dynamic of racialization operates through instrumental, hegemonic power relationships that also support heteronormativity (Butler, 2006), hegemonic masculinity and other social inequalities such as class. The intersectionality of inequalities acts differently according to the circumstances of time and place, but there is an essential loss of agency attributable to the ways in which different types of inequality act on each other. Historically, power has been held by elite white, heterosexual men from the West. That means lower-class white men can take some power and agency from their position in the social hierarchy (the power of being white, being male), but their class excludes them from real hegemonic power. Similarly, elite white women have power because of their higher social class and their whiteness, but their gender limits their freedoms. There is no need to get into a debate about which of these forms of instrumental control is more important: they are all ways in which people with power ensure that they keep power, and they are all ways in which the agency of others is limited. It would be almost possible for me to write a book on Masculinity and Leisure by simply removing whiteness from this text and replacing it with the word masculinity: both are products of history and both continue to act in shaping people's experiences of leisure. As I have shown throughout this book, it is impossible at times to separate out masculinity from whiteness, or elite status from whiteness. There is a triple helix of hegemonic power relationships that serve to legitimize and justify each other's interests, which are the interests of the same group of people: the white, elite class men who still have political, social and cultural hegemony today.

In leisure, as I have shown, instrumental whiteness constrains and limits the ways in which we might think of the transformative potential of leisure – as a space where communicative action might take precedence over instrumentality. Leisure could be an activity, a space, a moment in time where racialization is challenged, where hegemonies do not reach, where individuals could make free choices in a democratic, inclusive and social manner. Leisure has the potential to be subversive, transgressive, counter-cultural and a site of resistance and intentionality (Rojek, 2010). Leisure can be communicative and transformative, a way of constructing commonalities, sharing and creating new ways of being and new ways of becoming. In many ways, leisure always played this communicative role in making humans feel human and feel free from the daily struggles of life: leisure has meaning and purpose in its ability to console, to please and to engage (Spracklen, 2011). But the communicative value of leisure is increasingly swamped by the rising tide of instrumentality in this present century (Spracklen, 2009). Habermas is correct to observe the increasing dominance of instrumental rationality as the most dangerous shift in our late modernity because such rationality acts to reduce or even eliminate altogether our ability to be fully human: to follow Blumenberg's (2010) detours.

There is a pessimistic tone in the argument throughout this book, and so far in this conclusion. But it is important to note that the hegemony of instrumental whiteness is not a necessary condition of modernity, it is a contingent one. It is a pure accident of historical circumstances that has led to the rise of the West and the continued hegemonic dominance of the West. This hegemony is difficult to step around, and it feels as if it is a permanent and natural state of affairs: it feels like it has a necessary existence. But that enormity of presence and feeling of inevitability is itself a product of the way in which the hegemony tries to fool people into accepting the current circumstances. Just because instrumentality offers easy ways to give up thinking and give up questioning does not mean we have to accept those ways. There is no guarantee that the hegemony of the West, the dominance of elite white men, will continue into the second half of this century. Any number of possible futures awaits the world, assuming we manage to solve the pending environmental crises associated with climate change. Of those possible futures, three are probable. Firstly, Western hegemony may continue unchallenged and unsurpassed, in which case the commodification of leisure will be essentially complete, and the instrumental whiteness of leisure will continue unnoticed by anyone (except possibly the readers of this book) – this might be the most likely to occur, if we are to believe Habermas, though I am reluctant to make such a prediction. Secondly, Western hegemony may be replaced by another hegemony, possibly Chinese or perhaps some other new world power we cannot predict; on this future, leisure will function to promote the interests of this new hegemony, and although the nature of leisure activities will change to fit the new hegemony, the commodified and instrumental state of leisure will remain. Finally, we could use our communicative agency to radically reshape our political and social institutions; through communicative action, we could reject instrumental ways of thinking and try to create a world in which true equality operates to redistribute wealth, power and capital. In this lifeworld space, where the struggle to overturn hegemony begins, leisure might become again something truly communicative.

References

Achinstein, P. (1968) *Concepts of Science: A Philosophical Analysis* (Baltimore: Johns Hopkins).

Adorno, T. (1947) *Composing for the Films* (New York: Oxford University Press).

Adorno, T. (1991) *The Culture Industry* (London: Routledge).

Aitchison, C. (2001) 'Theorizing Other Discourses of Tourism, Gender and Culture: Can the Subaltern Speak (in Tourism)?', *Tourist Studies*, 1, 133–47.

Allison, P., Davis-Berman, J. and Berman, D. (2012) 'Changes in Latitude, Changes in Attitude: Analysis of the Effects of Reverse Culture Shock – A Study of Students Returning from Youth Expeditions', *Leisure Studies*, 31, 487–503.

Anderson, B. (1983) *Imagined Communities* (London: Verso).

Andrejevic, M. (2008) 'Watching Television without Pity: The Productivity of Online Fans', *Television and New Media*, 9, 24–46.

Anthias, F. (2003) 'Where Do I Belong? Narrating Collective Identity and Translocational Positionality', *Ethnicities*, 2, 491–514.

Appelrouth, S. (2011) 'Boundaries and Early Jazz: Defining a New Music', *Cultural Sociology*, 5, 225–242.

Arai, S. and Kivel, B.D. (2009) 'Critical Race Theory and Social Justice Perspectives on Whiteness, Difference(s) and (Anti)Racism: A Fourth Wave of Race Research in Leisure Studies', *Journal of Leisure Research*, 41, 459–470.

Archer, L. (2001) 'Muslim Brothers, Black Lads, Traditional Asians: British Muslim Young Men's Constructions of Race, Religion and Masculinity', *Feminism and Psychology*, 11, 79–105.

Archetti, E. (1995) 'In Search of National Identity: Argentinian Football and Europe', *International Journal of the History of Sport*, 12, 201–219.

Asante, M.K. (2001) 'Review of Paul Gilroy's *Against Race*', *Journal of Black Studies*, 31, 846–851.

Ashbee, E. (2011) 'Imperial Missions and the American State', *Journal of Political Power*, 4, 456–464.

Avila, E. (2006) *Popular Culture in the Age of White Flight: Fear and Fantasy in Suburban Los Angeles* (Berkeley: University of California Press).

Azarya, V. (2004) 'Globalization and International Tourism in Developing Countries: Marginality as a Commercial Commodity', *Current Sociology*, 52, 949–967.

Back, L., Crabbe, T. and Solomos, J. (2001) 'Lions and Black Skins: Race, Nation and Local Patriotism', in B. Carrington and I. McDonald (eds) *'Race', Sport and British Society* (London: Routledge).

Bairner, A. (2001) *Sport, Nationalism, and Globalization: European and North American Perspectives* (Albany: State University of New York Press).

Barker, C. (1999) *Television, Globalization and Cultural Identities* (Buckingham: Open University Press).

Bishop, H. and Jaworski, A. (2003) 'We Beat' em: Nationalism and the Hegemony of Homogeneity in the British Press Reportage of Germany versus England during Euro 2000', *Discourse and Society*, 14, 243–271.

Blackshaw, T. (2010) *Leisure* (London: Routledge).

Blair, M.E. (1993) 'Commercialization of the Rap Music Youth Subculture', *The Journal of Popular Culture*, 27, 21–33.

Blumenberg, H. (2010) *Care Crosses the River* (Stanford: Stanford University Press).

Bohlman, P. (2002) *World Music* (Oxford: Oxford University Press).

Bold, V. and Gillespie, S. (2009) 'The Southern Upland Way: Exploring Landscape and Culture', *International Journal of Heritage Studies*, 15, 245–257.

Bond, P. (2000) *Elite Transition: From Apartheid to Neoliberalism in South Africa* (London: Pluto Press).

Booth, D. (2003) 'Hitting Apartheid for Six? The Politics of the South African Sports Boycott', *Journal of Contemporary History*, 38, 477–493.

Booth, D. (2011) 'Olympic City Bidding: An Exegesis of Power', *International Review for the Sociology of Sport*, 46, 367–386.

Borradori, G. (2004) *Philosophy in a Time of Terror: Dialogues with Jurgen Habermas and Jacques Derrida* (Chicago: University of Chicago Press).

Borsay, P. (2005) *A History of Leisure* (Basingstoke: Palgrave Macmillan).

Bose, S. and Jalal, A. (2003) *Modern South Asia: History, Culture, Political Economy* (London: Routledge).

Bourdieu, P. (1986) *Distinction* (London: Routledge).

Bourdieu, P. (1998) *On Television and Journalism* (London: Pluto Press).

Braithwaite, R. (1962) 'Models in the Empirical Sciences', in E. Nagel, P. Suppes and A. Tarski (eds) *Logic, Methodology and the Philosophy of Science* (Stanford: Stanford University Press).

Bramham, P. (2006) 'Hard and Disappearing Work: Making Sense of the Leisure Project', *Leisure Studies*, 25, 379–390.

Brayton, S. (2005) ' "Black-Lash": Revisiting the "White Negro" through Skateboarding', *Sociology of Sport Journal*, 22, 356–372.

Breaux, R. (2010) 'After 75 Years of Magic: Disney Answers its Critics, Rewrites African American History, and Cashes in on its Racist Past', *Journal of African American Studies*, 14, 398–416.

Briggs, A. and Burke, P. (2009) *A Social History of The Media* (Cambridge: Polity).

Bruce, T. (2004) 'Marking the Boundaries of the "Normal" in Televised Sports: The Play-by-Play of Race', *Media, Culture and Society*, 26, 861–879.

Bryman, A. (2004) *The Disneyization of Society* (London: Sage).

Buettner, E. (2009) 'Chicken Tikka Masala, Flock Wallpaper, and "Real" Home Cooking: Assessing Britain's "Indian" Restaurant Traditions', *Food and History*, 7, 203–229.

Buford May, R. and Chaplin, K. (2008) 'Cracking the Code: Race, Class, and Access to Nightclubs in Urban America', *Qualitative Sociology*, 31, 57–72.

Burdsey, D. (2011) 'Strangers on the Shore? Racialized Representation, Identity and In/visibilities of Whiteness at the English Seaside', *Cultural Sociology*, 5, 537–552.

Burr, V. (2003) 'Ambiguity and Sexuality in Buffy the Vampire Slayer: A Sartrean Analysis', *Sexualities*, 6, 343–360.

Butler, J. (2006) *Gender Trouble: Feminism and the Subversion of Identity* (London: Routledge).

Butler, R. (2004) 'Geographical Research on Tourism, Recreation and Leisure: Origins, Eras and Directions', *Tourism Geographies*, 6, 143–162.

Carnap, R. (1939) *Foundations of Logic and Mathematics* (Chicago: University of Chicago Press).

Carnegie, E. and Mccabe, S. (2008) 'Re-enactment Events and Tourism: Meaning, Authenticity and Identity', *Current Issues in Tourism*, 11, 349–368.

Carrington, B. (1998) 'Sport, Masculinity and Black Cultural Resistance', *Journal of Sport and Social Issues*, 22, 275–298.

Carrington, B. (2004a) 'Introduction: Race/Nation/Sport', *Leisure Studies*, 23, 1–3.

Carrington, B. (2004b) 'Cosmopolitan Olympism, Humanism and the Spectacle of "Race"', in J. Bale and M.K. Cristensen (eds) *Post-Olympism? Questioning Sport in the Twenty-First Century* (Oxford: Berg).

Carrington, B. (2010a) 'What I said Was Racist – But I'm Not a Racist: Anti-Racism and the White Sports/Media Complex', in J. Long and K. Spracklen (eds) *Sport and Challenges to Racism* (Basingstoke: Palgrave Macmillan).

Carrington, B. (2010b) *Race, Sport and Politics: The Sporting Black Diaspora* (London: Sage).

Carrington, B. and McDonald, I. (eds) (2001a) *'Race' Sport and British Society* (London: Routledge).

Carrington, B. and McDonald, I. (2001b) 'Whose Game is it Anyway? Racism in Local League Cricket', in B. Carrington and I. McDonald (eds) *'Race', Sport and British Society* (London: Routledge).

Carrington, B. and McDonald, I. (2008) *Marxism, Cultural Studies and Sport* (London: Routledge).

Cartwright, N. (1983) *How the Laws of Physics Lie* (Oxford: Oxford University Press).

Cartwright, N. (1997) 'Models: The Blueprints for Laws', *Philosophy of Science*, 64, 292–303.

Castells, M. (1996) *The Rise of the Network Society* (Oxford: Blackwell).

Cheng, T.-M. and Tsaur, S.-H. (2012) 'The Relationship between Serious Leisure Characteristics and Recreation Involvement: A Case Study of Taiwan's Surfing Activities', *Leisure Studies*, 31, 53–68.

Clarke, S. and Garner, S. (2009) *White Identities: A Critical Sociological Approach* (London: Pluto Press).

Coalter, F. (1997) 'Leisure Sciences and Leisure Studies: Different Concept, Same Crisis?', *Leisure Sciences*, 19, 255–268.

Cochrane, J. (2008) 'Changing Landscapes and Rural Tourism', in R. Thomas (ed) *Managing Regional Tourism* (Ilkley: Great Northern Books).

Cohen, A.P. (1985) *The Symbolic Construction of Community* (London: Tavistock).

Cohen, S. (2010) 'Personal Identity (De)formation among Lifestyle Travellers: A Double-edged Sword', *Leisure Studies*, 29, 289–302.

Collins, T. (1999) *Rugby's Greatest Split* (London: Frank Cass).

Collins, T. and Vamplew, W. (2002) *Mud, Sweat and Beers: A Cultural History of Sport and Alcohol* (Oxford: Berg).

Connell, J. and Gibson, C. (2004) 'World Music: Deterritorializing Place and Identity', *Progress in Human Geography*, 28, 342–361.

Connell, R. (1995) *Masculinities* (Cambridge: Polity).

Cook, N. and Pople, A. (2004) *The Cambridge History of Twentieth-Century Music* (Cambridge: Cambridge University Press).

Cormack, B. (1998) *A History of Holidays, 1812–1990* (London: Routledge).

Corn, A. (2010) 'Land, Song, Constitution: Exploring Expressions of Ancestral Agency, Intercultural Diplomacy and Family Legacy in the Music of Yothu Yindi with Mandawuy Yunupinu', *Popular Music*, 29, 81–102.

Cosgrove, A. and Bruce, T. (2005) ' "The Way New Zealanders Would Like to See Themselves": Reading White Masculinity via Media Coverage of the Death of Sir Peter Blake', *Sociology of Sport Journal*, 22, 336–355.

Counihan, B. (2008) ' "When I Wake Up in the Morning, It All Depends on What I Want to Do": An Ethnography of Leisure in the Lives of Elderly Women', *Journal of Aging, Humanities, and the Arts*, 2, 25–35.

Coury, D. (2009) ' "Torn Country": Turkey and the West in Orhan Pamuk's *Snow*', *Critique: Studies in Contemporary Fiction*, 50, 340–349.

Crabbe, T. (2004): 'Englandfans – A New Club for a New England? Social Inclusion, Authenticity and the Performance of Englishness at Home and Away', *Leisure Studies*, 23, 63–78.

Cross, M. and Keith, M. (1993) *Racism, the City and the State* (London: Routledge).

Curran, J. and Seaton, J. (1991) *Power without Responsibility: The Press and Broadcasting in Britain* (Fourth Edition) (London: Routledge).

Darnell, S. (2007) 'Playing with Race: Right to Play and the Production of Whiteness in "Development through Sport" ', *Sport in Society*, 10, 560–579.

Dashper, K. (2010) ' "It's a Form of Freedom": The Experiences of People with Disabilities within Equestrian Sport', *Annals of Leisure Research*, 13, 86–101.

Dashper, K. (2012) ' "Dressage is Full of Queens!": Masculinity, Sexuality and Equestrian Sport', *Sociology*, published on-line at DOI: 10.1177/0038038512437898.

Davies, H. (2002) *Wainwright: The Biography* (London: Orion).

Daynes, S. and Lee, O. (2008) *Desire for Race* (Oxford: Oxford University Press).

Deakin, R. (2000) *Waterlog* (London: Vintage).

Debord, G. (1995[1967]) *The Society of the Spectacle* (London: Zone Books).

de Groot, J. (2006) 'Empathy and Enfranchisement: Popular Histories', *Rethinking History: The Journal of Theory and Practice*, 10, 391–413.

Delgado, R. and Stefancic, J. (eds) (1997) *Critical White Studies: Looking Behind the Mirror* (Philadelphia: Temple University Press).

den Breejen, L. (2007) 'The Experiences of Long Distance Walking: A Case Study of the West Highland Way in Scotland', *Tourism Management*, 28, 1417–1427.

Denison, J. and Markula, P. (2005) 'The Press Conference as a Performance: Representing Haile Gebrselassie', *Sociology of Sport Journal*, 22, 311–335.

Denzin, N. (2002) *Reading Race* (London: Sage).

Dimeo, P. and Finn, G. (2001) 'Racism, National Identity and Scottish Football', in B. Carrington and I. McDonald (eds) *'Race', Sport and British Society* (London: Routledge).

Doane, A. and Bonilla-Silva, E. (eds) (2003) *White Out: The Continuing Significance of Racism* (New York: Routledge).

Douglas, D. (2005) 'Venus, Serena, and the Women's Tennis Association: When and Where "Race" Enters', *Sociology of Sport Journal*, 22, 256–282.

Du Bois, W.E.B. (1986) *W.E.B. Du Bois: Writings* (New York: Library of America).

Dudrah, R. (2011) 'British Bhangra Music as Soundscapes of the Midlands', *Midland History*, 36, 278–291.

Dunning, E and Sheard, K. (2005) *Barbarians, Gentlemen and Players* (Second Edition) (London: Routledge).

Dyer, R. (1997) *White* (New York: Routledge).

Edensor, T. (2002) *National Identity, Popular Culture and Everyday Life* (Oxford: Berg).

Edensor, T. (2004) 'Automobility and National Identity: Representation, Geography and Driving Practice', *Theory, Culture and Society*, 21, 101–120.

Edensor, T. (2006) 'Reconsidering National Temporalities Institutional Times, Everyday Routines, Serial Spaces and Synchronicities', *European Journal of Social Theory*, 9, 525–545.

Edensor, T. and Richards, S. (2007) 'Snowboarders vs Skiers: Contested Choreographies of the Slopes', *Leisure Studies*, 26, 97–114.

Edwards, K. (2009) 'Good Looks and Sex Symbols: The Power of the Gaze and the Displacement of the Erotic in Twilight', *Screen Education*, 53, 26–32.

Ellis, S. (2011) 'Music Camp: Experiential Consumption in a Guitar Workshop Setting', *International Journal of Culture, Tourism and Hospitality Research*, 5, 76–382.

Emerson, R. (2002) '"Where My Girls At?": Negotiating Black Womanhood in Music Videos', *Gender and Society*, 16, 115–135.

Entine, J. (2000) *Taboo: Why Black Athletes Dominate Sports and Why We Are Afraid To Talk about It* (New York: Public Affairs).

Erickson, B. (2005) 'Style Matters: Explorations of Bodies, Whiteness, and Identity in Rock Climbing', *Sociology of Sport Journal*, 22, 373–396.

Erickson, B., Johnson, C. and Kivel, B.D. (2009) 'Rocky Mountain National Park: History and Culture as Factors in African-American Park Visitation', *Journal of Leisure Research*, 41, 529–545.

Falconer, R. and Kingham, S. (2007) 'Driving People Crazy: A Geography of Boy Racers in Christchurch, New Zealand', *New Zealand Geographer*, 63, 181–191.

Fanon, F. (1967) *Black Skin, White Masks* (New York: Grove Press).

Farred, G. (2004) 'Fiaca and Veronismo: Race and Silence in Argentine Football', *Leisure Studies*, 23, 47–61.

Fishkin, S.F. (1995) 'Interrogating Whiteness, Complicating Blackness: Remapping American Culture', *American Quarterly*, 47, 428–466.

Fishwick, M. (1952) 'The Cowboy: America's Contribution to the World's Mythology', *Western Folklore*, 11, 77–92.

Fiske, J. and Hartley, J. (2003) *Reading Television* (London: Psychology Press).

Fiske, S. (1993) 'Controlling Other People: The Impact of Power on Stereotyping', *American Psychologist*, 48, 621–628.

Fleming, S. (1991) 'Sport, Solidarity and Asian Male Youth Culture', in G. Jarvie (ed) *Sport, Racism and Ethnicity* (London: Falmer Press).

Fleming, S. (1994) 'Sport and South Asian Youth: The Perils of "False Universalism" and Stereotyping', *Leisure Studies*, 13, 159–177.

Fleming, S. (1995) *Home and Away: Sport and South Asian Youth* (Aldershot: Avebury).

Fleming, S. (2001) 'Racial Science and South Asian and Black Physicality', in B. Carrington and I. McDonald (eds) *'Race', Sport and British Society* (London: Routledge).

Floyd, M.F. (1998) 'Getting Beyond Marginality and Ethnicity: The Challenge For Race and Ethnic Studies in Leisure Research', *Journal of Leisure Research*, 30, 3–22.

Foucault, M. (1970) *The Order of Things* (London: Tavistock).

Foucault, M. (1972) *The Archaeology of Knowledge* (London: Tavistock).

Foucault, M. (1973) *The Birth of the Clinic* (London: Tavistock).

Foucault, M. (2006) *The History of Madness* (London: Routledge).

Fox, A. (1992) 'The Jukebox of History: Narratives of Loss and Desire in the Discourse of Country Music', *Popular Music*, 11, 53–72.

Frankenberg, R. (1994) *White Women, Race Matters* (Madison: University of Wisconsin Press).

Frankenberg, R. (ed) (1997) *Displacing Whiteness: Essays in Social and Cultural Criticism* (Durham: Duke University Press).

Frigyesi, J. (1994) 'Béla Bartók and the Concept of Nation and "Volk" in Modern Hungary', *The Musical Quarterly*, 78, 255–287.

Frith, S. (1998) *Performing Rites: On the Value of Popular Music* (Cambridge: Harvard University Press).

Fukuyama, F. (1992) *The End of History and the Last Man* (Harmondsworth: Penguin).

Fulton, G. and Bairner, A. (2007) 'Sport, Space and National Identity in Ireland: The GAA, Croke Park and Rule 42', *Space and Polity*, 11, 55–74.

Fusco, C. (2005) 'Cultural Landscapes of Purification: Sports Spaces and Discourses of Whiteness', *Sociology of Sport Journal*, 22, 283–310.

Gabriel, J. (1998) *Whitewash: Racialized Politics and the Media* (London: Routledge).

Gardiner, S. and Welch, R. (2001) 'Sport, Racism and the Limits of "Color Blind" Law', in B. Carrington and I. McDonald (eds) *'Race', Sport and British Society* (London: Routledge).

Garland, J. (2004) 'The Same Old Story? Englishness, the Tabloid Press and the 2002 Football World Cup', *Leisure Studies*, 23, 79–92.

Garner, S. (2003) *Racism in the Irish Experience* (London: Pluto Press).

Garner, S. (2006) 'The Uses of Whiteness: What Sociologists in Europe Can Draw From US Research on Whiteness', *Sociology*, 40, 257–275.

Garner, S. (2007) *Whiteness: An Introduction* (London: Routledge).

Gemie, S. (2005) 'Roots, Rock, Breizh: Music and the Politics of Nationhood in Contemporary Brittany', *Nations and Nationalism*, 11, 103–120.

Gibson, H. (1998) 'Active Sport Tourism: Who Participates?', *Leisure Studies*, 17, 155–170.

Giere, R. (1988) *Explaining Science* (Chicago: University of Chicago Press).

Giere, R. (1999) *Science without Laws* (Chicago: University of Chicago Press).

Gill, R. (2006) *Gender and the Media* (Cambridge: Polity).

Gilman, S. (1985) *Difference and Pathology: Stereotypes of Sexuality, Race and Madness* (London: Cornell).

Gilroy, P. (1987) *There Ain't no Black in the Union Jack* (London: Routledge).

Gilroy, P. (1992) *The Black Atlantic: Modernity and Double Consciousness* (London: Verso).

Gilroy, P. (2000) *Between Camps: Nations, Culture and the Allure of Race* (London: Allen Lane).

Goldberg, D. (1993) *Racist Culture* (Oxford: Blackwell).

Gonzalez, S. (2010) 'Bilbao and Barcelona in Motion: How Urban Regeneration Models Travel and Mutate in the Global Flows of Policy Tourism', *Urban Studies*, 48, 1397–1418.

Gordon, W., Miller, F. and Rollock, D. (1990) 'Coping with Communicentric Bias in Knowledge Production', *Educational Researcher*, 19, 14–19.

Gotham, K. (2002) 'Marketing Mardi Gras: Commodification, Spectacle and the Political Economy of New Orleans', *Urban Studies*, 39, 1735–796.

Gournelos, E. (2009) 'Blasphemous Allusion: Coming of Age in South Park', *Journal of Communication Inquiry*, 33, 143–168.

Grainger, A., Falcous M. and Jackson, J. (2012) 'Postcolonial Anxieties and the Browning of New Zealand Rugby', *The Contemporary Pacific*, 24, 267–295.

Grant, E. (2005) 'Race and Tourism in America's First City', *Journal of Urban History*, 31, 850–871.

Green, E. and Singleton, C. (2006) 'Risky Bodies at Leisure: Young Women Negotiating Space and Place', *Sociology*, 40, 853–871.

Gupta, S. (2009) *Re-reading Harry Potter* (Second Edition) (Basingstoke: Palgrave).

Guralnick, P. (1995) *Last Train to Memphis* (London: Abacus).

Habermas, J. (1981) 'Modernity versus Postmodernity', *New German Critique*, 22, 3–14.

Habermas, J. (1984) *The Theory of Communicative Action, Volume One: Reason and the Rationalization of Society* (Cambridge: Polity).

Habermas, J. (1987) *The Theory of Communicative Action, Volume Two: The Critique of Functionalist Reason* (Cambridge: Polity).

Habermas, J. (1989[1962]) *The Structural Transformation of the Public Sphere* (Cambridge: Polity).

Habermas, J. (1990) *The Philosophical Discourse of Modernity* (Cambridge: Polity).

Habermas, J. (2000) *Post-National Constellation* (Cambridge: Polity).

Hage, G. (1998) *White Nation: Fantasies of White Supremacy in a Multicultural Society* (Annandale: Pluto).

Hahn, L.E. (2000) *Perspectives on Habermas* (Illinois: Open Court).

Halberstam, J. (1993) 'Technologies of Monstrosity: Bram Stoker's Dracula', *Victorian Studies* 36, 333–352.

Hall, R. (2001) 'The Ball Curve: Calculated Racism and the Stereotype of African American Men', *Journal of Black Studies*, 32, 104–119.

Hall, S. (1993) 'Culture, Community, Nation', *Cultural Studies*, 7, 349–363.

Hall, S. (1995) 'Negotiating Caribbean Identities', *New Left Review*, 209, 3–14.

Hall, S. (1996) *Modernity: An Introduction to Modern Societies* (London: Blackwell).

Hall, S. (1997) 'Subjects in History: Making Diasporic Identities', in W. Lubiani (ed) *The House that Race Built* (New York: Pantheon).

Hall, S. (2002) 'Race, Articulation, and Societies Structured in Dominance', in P. Essed and D. Goldberg (eds) *Race Critical Theories* (London: Blackwell).

Hargrove, M. (2009) 'Mapping the "Social Field of Whiteness": White Racism as Habitus on the City where History Lives', *Transforming Anthropology*, 17, 93–104.

Hark, I.R. (2008) *Star Trek* (Basingstoke: Palgrave).

Haynes, R. (2009) 'Lobby and the Formative Years of Radio Sports Commentary, 1935–1952', *Sport in History*, 29, 25–48.

Hebdige, D. (1979) *Subcultures: The Meaning of Style* (London: Routledge).

Hefner, B.E. (2012) 'Rethinking *Blacula*: Ideological Critique at the Intersection of Genres', *Journal of Popular Film and Television*, 40, 62–74.

Hegarty, P. (2008) 'Constructing (in) the 'Real' World: Simulation and Architecture in Baudrillard', *French Cultural Studies*, 19, 317–331.

Henderson, K. (1998) Researching Diverse Populations', *Journal of Leisure Research*, 30, 157–174.

Henderson, K. and Ainsworth, B. (2001) 'Researching Leisure and Physical Activity with Women of Color: Issues and Emerging Questions', *Leisure Sciences*, 23, 21–34.

Henderson, K., Presley, J. and Bialeschki, M.D. (2004) 'Theory in Recreation and Leisure Research: Reflections from the Editors', *Leisure Sciences*, 26, 411–425.

Hesse, M. (1963) *Models and Analogies in Science* (Oxford: Oxford University Press).

Heywood, I. (2006) 'Climbing Monsters: Excess and Restraint in Contemporary Rock Climbing', *Leisure Studies*, 25, 455–467.

Hiller, H. and Wanner, R. (2011) 'Public Opinion in Host Olympic Cities: The Case of the 2010 Vancouver Winter Games', *Sociology*, 45, 883–899.

Hobsbawm, E. (1992) *Nations and Nationalism since 1780* (Cambridge: Cambridge University Press).

Hollands, R. and Chatterton, P. (2002) 'Changing Times for an Industrial City', *City*, 6, 291–315.

Holmes, D. (2001) *Virtual Globalization: Virtual Spaces/Tourist Spaces* (London: Psychology Press).

Hope, W. (2002) 'Whose All Blacks?', *Media, Culture and Society*, 24, 235–253.

Horan, B. (1988) 'Theoretical Models, Biological Complexity and the Semantic View of Theories', *Philosophy of Science*, 2, 265–277.

Horne, J. (2006) *Sport in Consumer Culture* (Basingstoke: Palgrave Macmillan).

Hugenberg, L., Haridakis, P. and Earnheardt, A. (2008) *Sports Mania: Essays on Fandom in the Twenty-First Century* (Jefferson: McFarland and Co).

Hughey, M. and Muradi, S. (2009) 'Laughing Matters: Economies of Hyper-Irony and Manic-Satire in South Park and Family Guy', *Humanity and Society*, 33, 206–237.

Hylton, K. (2005) ' "Race", Sport and Leisure: Lessons from Critical Race Theory', *Leisure Studies*, 24, 81–98.

Hylton, K. (2009) *'Race' and Sport: Critical Race Theory* (London: Routledge).

Iqani, M. (2012) 'Smooth Bodywork: The Role of Texture on Images of Cars and Women on Consumer Magazine Covers', *Social Semiotics*, 22, 311–331.

Jaher, F.C. (2001) 'Antisemitism in American Athletics', *Shofar: An Interdisciplinary Journal of Jewish Studies*, 20, 61–73.

Jayne, M., Valentine, G. and Holloway, S. (2008) 'Fluid Boundaries – British Binge Drinking and European Civility: Alcohol and the Production and Consumption of Public Space', *Space and Polity*, 12, 81–100.

Jenkins, J. and James, P. (1994) *From Acorn to Oak Tree: The Growth of the National Trust 1895–1914* (London: Macmillan).

Jennings, A. (2011) 'Investigating Corruption in Corporate Sport: The IOC and FIFA,' *International Review for the Sociology of Sport*, 46, 387–398.

Jerrome, D. (1984) 'Good Company: The Sociological Implications of Friendship', *The Sociological Review*, 32, 696–718.

Johal, S. (2001) 'Playing their Own Game: A South Asian Football Experience', in B. Carrington and I. McDonald (eds) *'Race', Sport and British Society* (London: Routledge).

Jones, O. (2011) *Chavs: The Demonization of the Working Class* (London: Verso).

Kassimeris, C. (2008) *European Football in Black and White: Tackling Racism in Football* (Lanham: Lexington Books).

Kellner, D. (1995) *Cultural Studies, Identity and Politics between the Modern and the Post-Modern* (London: Psychology Press).

Keys, B. (2004) 'Spreading Peace, Democracy, and Coca-Cola', *Diplomatic History*, 28, 165–196.

Kidd, D. (2007) 'Harry Potter and the Functions of Popular Culture', *The Journal of Popular Culture*, 40, 69–89.

King, C. (2004) 'Race and Cultural Identity: Playing the Race Game inside Football', *Leisure Studies*, 23, 19–30.

King, C. (2005) 'Cautionary Notes on Whiteness and Sport Studies', *Sociology of Sport Journal*, 22, 397–408.

King, C. (2007) 'Staging the Winter *White Olympics* Or, Why Sport Matters to *White* Power', *Journal of Sport and Social Issues*, 31, 89–94.

Kirk, N. (2000) *Northern Identities: Historical Interpretations of the North and Northerness* (Aldershot: Ashgate).

Kivel, B.D., Johnson, C. and Scraton, S. (2009) '(Re)Theorizing Leisure, Experience and Race', *Journal of Leisure Research*, 41, 473–493.

Klein, N. (2000) *No Logo* (London: Harper Collins).

Klemm, M. (2002) 'Tourism and Ethnic Minorities in Bradford: The Invisible Segment', *Journal of Travel Research*, 41, 85–91.

Kohn, M. (1995) *The Race Gallery* (London: Verso).

Kusz, K. (2001) 'I Want to be the Minority: The Politics of Youthful White Masculinities in Sport and Popular Culture in 1990s America', *Journal of Sport and Social Issues*, 25, 390–416.

Lashua, B. (2007) 'Making an Album: Rap Performance and a CD Track Listing as Performance Writing in *The Beat of Boyle Street* Music Programme', *Leisure Studies*, 26, 429–445.

Lee, C. (2012) 'Have Magic, Will Travel: Tourism and Harry Potter's United (Magical) Kingdom', *Tourist Studies*, 12, 52–69.

Lefebvre, H. (1991[1947]) *Critique of Everyday Life* (London: Verso).

Lindsey, E. (2001) 'Notes from the Sports Desk: Reflections on Race, Class and Gender in British Sports Journalism', in B. Carrington and I. McDonald (eds) *'Race', Sport and British Society* (London: Routledge).

Lipsitz, G. (1990) 'Listening to Learn and Learning to Listen: Popular Culture, Cultural Theory, and American Studies', *American Quarterly*, 42, 615–636.

Littlefield, J. and Siudzinski, R. (2012) ' "Hike your own Hike": Equipment and Serious Leisure along the Appalachian Trail', *Leisure Studies*, 31, 465–486.

Lloyd, E. (1984) 'A Semantic Approach to the Structure of Population Genetics', *Philosophy of Science*, 51, 242–264.

Lomax, M. (1998) 'If He were White: Portrayals of Black and Cuban Players in Organized Baseball, 1880–1920', *Journal of African American Studies*, 3, 31–44.

Long, J. and Hylton, K. (2002) 'Shades of White: An Examination of Whiteness in Sport', *Leisure Studies*, 16, 87–103.

Long, J. and Spracklen, K. (eds) (2010) *Sport and Challenges to Racism* (Basingstoke: Palgrave Macmillan).

Long, J., Carrington, B., and Spracklen, K. (1997) ' "Asians Cannot Wear Turbans in the Scrum": Explorations of Racist Discourse Within Professional Rugby League', *Leisure Studies*, 16, 249–260.

Long, J., Robinson, P. and Spracklen, K. (2005) 'Promoting Racial Equality within Sports Organizations', *Journal of Sport and Social Issues*, 29, 41–59.

Lucas, C., Deeks, M. and Spracklen, K. (2011) 'Grim Up North: Northern England, Northern Europe and Black Metal', *Journal for Cultural Research*, 15, 279–296.

Lyons, A. and Willott, S. (2008) 'Alcohol Consumption, Gender Identities and Women's Changing Social Positions', *Sex Roles*, 59, 694–712.

MacCannell, D. (1973) 'Staged Authenticity: Arrangements of Social Space in Tourist Settings', *American Journal of Sociology*, 79, 589–603.

Machin, A. (2008) 'A History of Tourism in Yorkshire', in R. Thomas (ed) *Managing Regional Tourism* (Ilkley: Great Northern Books).

Mahon, M. (2000) 'Black Like This: Race, Generation, and Rock in the Post-Civil Rights Era', *American Ethnologist*, 27, 283–311.

Maira, S. (2008) 'Belly Dancing: Arab-Face, Orientalist Feminism, and US Empire', *American Quarterly*, 60, 317–345.

Mangan, J. (1981) *Athleticism in the Victorian and Edwardian Public Schools* (Cambridge: Cambridge University Press).

Mangan, J. (1986) *The Games Ethic and Imperialism: Aspects of the Diffusion of an Ideal* (London: Frank Cass).

Mangan, J. (1995) 'Duty unto death: English Masculinity and Militarism in the Age of the New Imperialism', in J. Mangan (ed) *Tribal Identities: Nationalism, Europe, Sport* (London: Frank Cass).

Mangan, J. and Ritchie, A. (2005) *Ethnicity, Sport, Identity: Struggles for Status* (London: Routledge).

Marqusee, M. (2001) 'In Search of the Unequivocal Englishman: The Conundrum of Race and Nation in English Cricket', in B. Carrington and I. McDonald (eds) *'Race', Sport and British Society* (London: Routledge).

Matheson, C. (2008) 'Music, Emotion and Authenticity: A Study of Celtic Music Festival Consumers', *Journal of Tourism and Cultural Change*, 6, 57–74.

Mayer, V. (2001) 'Pop Goes the World', *Emergences: Journal for the Study of Media and Composite Cultures*, 11, 309–324.

McDonald, M. (2005) 'Mapping Whiteness and Sport: An Introduction', *Sociology of Sport Journal*, 22, 245–255.

McDonald, M. (2009) 'Dialogues on Whiteness, leisure, and (Anti)Racism', *Journal of Leisure Research*, 41, 5–21.

McGuigan, J. (2006) *Modernity and Postmodern Culture* (Maidenhead: Open University Press).

Meehan, J. (1995) *Feminists Read Habermas: Gendering the Subject of Discourse* (London: Routledge).

Merriman, P. (2005) 'Respect the Life of the Countryside: The Country Code, Government and the Conduct of Visitors to the Countryside in Post-War England and Wales', *Transactions of the Institute of British Geographers*, 30, 336–350.

Miles, R. (1989) *Racism* (London: Routledge).

Miles, R. (1993) *Racism after 'Race Relations'* (London: Routledge).

Millward, P. (2011) *The Global Football League: Transnational Networks, Social Movements and Sport in the New Media Age* (Basingstoke: Palgrave Macmillan).

Mitchell, W. (2009) *Mitchell: From Milltown to Mountain* (Ilkley: Great Northern Books).

Mitter, R. (2004) *A Bitter Revolution: China's Struggle with the Modern World* (Oxford: Oxford University Press).

Moore, A. (2000) 'Opera of the Proletariat: Rugby League, the Labour Movement and Working-Class Culture in New South Wales and Queensland', *Labour History*, 79, 57–70.

Morgan, W. (2006) *Why Sports Morally Matter* (London: Routledge).

Morrison, M. (1999) 'Models as Autonomous Agents', in M. Morgan and M. Morrison (eds) *Models as Mediators* (Cambridge: Cambridge University Press).

Morrow, S. (2005) 'Continuity and Change: The Planning and Management of Long Distance Walking Routes in Scotland', *Managing Leisure*, 10, 237–250.

Mowatt, R. (2009) 'Notes from a Leisure Son: Expanding an Understanding of Whiteness in Leisure', *Journal of Leisure Research*, 41, 511–528.

Munt, I. (1994) 'Eco-tourism or Ego-tourism?', *Race and Class*, 36, 49–60.

Nanton, P. (1989) 'The New Orthodoxy: Racial Categories and Equal Opportunity Policy', *New Community*, 15, 549–564.

Nauright, J. (1996) 'A Besieged Tribe? Nostalgia, White Cultural Identity and the Role of Rugby in a changing South Africa', *International Review for the Sociology of Sport*, 31, 69–86.

Nayak, A. (2003) *Race, Place and Globalisation: Youth Cultures in a Changing World* (Oxford: Berg).

Nayak, A. (2006) 'After Race: Ethnography, Race and Post-Race Theory', *Ethnic and Racial Studies*, 29, 411–430.

Nebeker, K. (1998) 'Critical Race Theory: A White Graduate Student's Struggle with this Growing Area of Scholarship', *Qualitative Studies in Education*, 11, 25–41.

Obrador, P. (2011) 'The Place of the Family in Tourism Research: Domesticity and Thick Sociality by the Pool', *Annals of Tourism Research*, 39, 401–420.

O'Connell Davidson, J. (1996) 'Sex Tourism in Cuba', *Race and Class*, 38, 39–48.

Oliver, P. (1990) *Blues Fell This Morning* (Cambridge: Cambridge University Press).

Olsen, W., Warde, A. and Martens, L. (2000) 'Social Differentiation and the Market for Eating Out in the UK', *International Journal of Hospitality Management*, 19, 173–190.

Omi, M. and Winant, H. (1986) *Racial Formation in the United States* (New York: Routledge).

Opotow, S. (1990) 'Moral Exclusion and Injustice: An Introduction', *Journal of Social Sciences*, 46, 1–20.

O'Sullivan, E. (2003) 'Bringing a Perspective of Transformative Learning to Globalized Consumption', *International Journal of Consumer Studies*, 27, 326–330.

Palmer, C. and Brady, E. (2007) 'Landscape and Value in the Work of Alfred Wainwright (1907–1991)', *Landscape Research*, 32, 397–421.

Pamuk, O. (2004) *Snow* (London: Faber and Faber).

Papacharissi, Z. (2011) *A Networked Self: Identity, Community, and Culture on Social Network Sites* (London: Routledge).

Parekh, B. (2000) *Re-thinking Multiculturalism* (Basingstoke: Palgrave Macmillan).

Parker, M. (2011) 'Organizing the Circus: The Engineering of Miracles', *Organization Studies*, 32, 555–569.

Parlebas, P. (2003) 'The Destiny of Games Heritage and Lineage', *Studies in Physical Culture and Tourism*, 10, 15–26.

Philipp, S. (1994) 'Race and Tourism Choice: A Legacy of Discrimination?', *Annals of Tourism Research*, 21, 479–488.

Philipp, S. (1995) 'Race and Leisure Constraints', *Leisure Sciences*, 17, 109–120.

Pierre, J. (2009) 'Beyond Heritage Tourism: Race and the Politics of African-Diasporic Interactions', *Social Text*, 27, 59–81.

Ratna, A. (2010) 'Taking the Power Back: The Politics of British-Asian Female Football Players, *Young*, 18, 117–132.

Rawls, J. (1971) *A Theory of Justice* (New York: Routledge).

Ray, K. and Srinivas, T. (2012) *Curried Cultures: Globalization, Food, and South Asia* (Berkeley: University of California Press).

Richmond, L. and Johnson, C. (2009) ' "It's a Race War:" Race and Leisure Experiences in California State Prison', *Journal of Leisure Research*, 41, 565–580.

Rivers-Moore, M. (2007) 'No Artificial Ingredients? Gender, Race and Nation in Costa Rica's International Tourism Campaign', *Journal of Latin American Cultural Studies: Travesia*, 16, 341–357.

Roberts, K. (2004) *The Leisure Industries* (Basingstoke: Palgrave).

Roberts, K. (2011) 'Leisure: The Importance of Being Inconsequential', *Leisure Studies* 30, 5–20.

Roberts, N. (2009) 'Crossing the Color Line With a Different Perspective on Whiteness and (Anti)Racism: A Response to Mary McDonald', *Journal of Leisure Research*, 41, 495–509.

Robson, E. (2003) *After Wainwright* (Wasdale: Striding Edge).

Rodriguez, S. (2001) 'Tourism, Whiteness and the Vanishing Anglo', in D. Wrobel and P. Long (eds) *Seeing and Being Seen: Tourism in the American West* (Lawrence: University Press of Kansas).

Roessingh, C. and Duijnhoven, H. (2005) 'Small Entrepreneurs and Shifting Identities: The Case of Tourism in Puerto Plata (Northern Dominican Republic)', *Journal of Tourism and Cultural Change*, 2, 185–202.

Rojek, C. (2010) *The Labour of Leisure* (London: Sage).

Rojek, C. and Urry, J. (1997) *Touring Cultures* (London: Routledge).

Rowe, D. (2003) 'Sport and the Repudiation of the Global', *International Review for the Sociology of Sport*, 38, 281–294.

Rowland, I. (2002) *The Full Facts Book About Cold Reading* (London: IR Publications).

Rowling, J.K. (1997) *Harry Potter and the Philosopher's Stone* (London: Bloomsbury).

Rudd, M, and Davis, J. (1998) 'Industrial Heritage Tourism at the Bingham Canyon Copper Mine', *Journal of Travel Research*, 36, 85–89.

Sack, A. and Suster, Z. (2000) 'Soccer and Croatian Nationalism: A Prelude to War', *Journal of Sport and Social Issues*, 24, 305–320.

Saeki, T. (1994) 'The Conflict between Tradition and Modernization in a Sport Organization: A Sociological Study of Issues Surrounding the Organizational Reformation of the all Japan Judo Federation', *International Review for the Sociology of Sport*, 29, 301–315.

Said, E. (1978) *Orientalism* (New York: Vintage).

Said, E. (1985) 'Orientalism Reconsidered', *Cultural Critique*, 1, 89–107.

Saldanha, A. (2005) 'Trance and Visibility at Dawn: Racial Dynamics in Goa's Rave Scene', *Social and Cultural Geography*, 6, 707–721.

Saville, S. (2008) 'Playing with Fear: Parkour and the Mobility of Emotion', *Social and Cultural Geography*, 9, 891–914.

Scambler, G. (2005) *Sport and Society: History, Power and Culture* (Maidenhead: Open University Press).

Scherer, J. and Sam, M. (2012) 'Public Broadcasting, Sport and Cultural Citizenship: Sky's the Limit in New Zealand?', *Media, Culture and Society*, 34, 101–111.

Scherer, J., Falcous, M. and Jackson, S. (2008) 'The Media Sports Cultural Complex: Local-Global Disjuncture in New Zealand/Aotearoa', *Journal of Sport and Social Issues*, 32, 48–71.

Scott, D. and Becken, S. (2010) 'Adapting to Climate Change and Climate Policy: Progress, Problems and Potentials, *Journal of Sustainable Tourism*, 18, 283–295.

Scraton, S. (2001) 'Reconceptualizing Race, Gender and Sport: The Contribution of Black Feminism', in B. Carrington and I. McDonald (eds) *'Race', Sport and British Society* (London: Routledge).

Searle, C. (2001) 'Pitch of Life: Re-reading CLR James' *Beyond a Boundary'*, in B. Carrington and I. McDonald (eds) *'Race', Sport and British Society* (London: Routledge).

Sen, C.T. (2009) *Curry: A Global History* (London: Reaktion).

Sheridan, L. (2006) 'Islamophobia Pre- and Post-September 11th, 2001', *Journal of Interpersonal Violence*, 21, 317–336.

Shinew, K.J., Floyd, M.F. and Parry, P. (2004) 'Understanding the Relationship between Race and Leisure Activities and Constraints: Exploring an Alternative Framework', *Leisure Sciences*, 26, 181–199.

Silverstein, P. (2004) *Algeria in France: Transpolitics, Race, Nation* (Bloomington: University of Indiana Press).

Silverstein, P. (2008) 'The Context of Antisemitism and Islamophobia in France', *Patterns of Prejudice*, 42, 1–26.

Silvia, T. (2007) *Baseball over the Air: The National Pastime on the Radio and in the Imagination* (Jefferson: McFarland).

Skinner, R. (2010) 'Civil Taxis and Wild Trucks: The Dialectics of Social Space and Subjectivity in Dimanche à Bamako', *Popular Music*, 29, 17–39.

Snape, R. (2004) 'The Co-operative Holidays Association and the Cultural Formation of Countryside Leisure Practice', *Leisure Studies*, 23, 143–158.

Solorzano, D. and Yosso, T. (2001) 'Critical Race Theory and Method: Counter-storytelling', *Qualitative Studies in Education*, 14, 471–495.

Spracklen, K. (2001) 'Black Pearl, Black Diamonds: Exploring Racial Identities in Rugby League', in B. Carrington and I. McDonald (eds) *'Race', Sport and British Society* (London: Routledge).

Spracklen, K. (2006) 'Leisure, Consumption and a Blaze in the Northern Sky: Developing an Understanding of Leisure at the End of Modernity through the Habermasian Framework of Communicative and Instrumental Rationality', *World Leisure Journal*, 48, 33–44.

Spracklen, K. (2008) 'The Holy Blood and the Holy Grail: Myths of Scientific Racism and the Pursuit of Excellence in Sport', *Leisure Studies*, 27, 221–227.

Spracklen, K. (2009) *The Meaning and Purpose of Leisure* (Basingstoke: Palgrave Macmillan).

Spracklen, K. (2011) *Constructing Leisure* (Basingstoke: Palgrave Macmillan).

Spracklen, K. (2012) 'Nazi Punks Folk Off: Leisure, Nationalism, Cultural Identity and the Consumption of Metal and Folk Music', *Leisure Studies*, published on-line at DOI:10.1080/02614367.2012.674152.

Spracklen, K. and Spracklen, B. (2012) 'Pagans and Satan and Goths, oh my: Dark Leisure as Communicative Agency and Communal Identity on the Fringes of the Modern Goth Scene', *World Leisure Journal*, 54, 350–362.

Spracklen, K., Long, J. and Hylton, K. (2006) 'Managing and Monitoring Equality and Diversity in UK Sport', *Journal of Sport and Social Issues*, 30, 289–305.

Spracklen, K., Timmins, S. and Long, J. (2010) 'Ethnographies of the Imagined, the Imaginary and the Critically Real: Blackness, Whiteness, the North of England and Rugby League', *Leisure Studies*, 29, 397–414.

Stebbins, R. (2013) *The Committed Reader* (Lanham: Scarecrow Press).

Stephenson, T. (1989) *Forbidden Land: The Struggle for Access to Mountain and Moorland* (Manchester: Manchester University Press).

St. Louis, B. (2004) 'Sport and Common-Sense Racial Science', *Leisure Studies*, 23, 31–46.

Strutt, J. (1801) *The Sports and Pastimes of the People of England from the Earliest Period* (Re-published 1903, London: Methuen and Company).

Suppe, F. (1977) 'The Search for Philosophic Understanding of Scientific Theories', in F. Suppe (ed) *The Structure of Scientific Theories* (Second Edition) (Urbana: University of Illinois Press).

Suppes, P. (1957) *Introduction to Logic* (Princeton: Van Nostrand).

Swinney, A. and Horne, J. (2005) 'Race Equality and Leisure Policy Discourses in Scottish Local Authorities', *Leisure Studies*, 24, 271–289.

Tatz, C. (2009) 'Coming to Terms: "Race", Ethnicity, Identity and Aboriginality in Sport', *Australian Aboriginal Studies*, 2, 15–31.

Thomas, H. and Piccolo, F. (2000) 'Best Value, Planning and Racial Equality', *Planning Practice and Research*, 15, 79–95.

Thomas, R. (ed) (2008) *Managing Regional Tourism* (Ilkley: Great Northern Books).

Thorpe, H. (2012) 'Sex, Drugs and Snowboarding: (Il)legitimate Definitions of Taste and Lifestyle in a Physical Youth Culture, *Leisure Studies*, 31, 33–51.

Urquia, N. (2005) 'The Re-Branding of Salsa in London's Dance Clubs: How an Ethnicised Form of Cultural Capital was Institutionalised', *Leisure Studies*, 24, 385–397.

Vamplew, W. (2004) *Pay Up and Play the Game: Professional Sport in Britain, 1875–1914* (Cambridge: Cambridge University Press).

Van Fraassen, B. (1980) *The Scientific Image* (Oxford: Clarendon).

Vincent, J., Kian, E., Pedersen, P., Kuntz, A. and Hill. J. (2010) 'England Expects: English Newspapers' Narratives about the English Football Team in the 2006 World Cup', *International Review for the Sociology of Sport*, 45, 199–223.

Wai-Chung, H. (2004) 'A Comparative Study of Music Education in Shanghai and Taipei: Westernization and Nationalization', *Compare: A Journal of Comparative and International Education*, 34, 231–249.

Wainwright, A. (1955) *A Pictorial Guide to the Lakeland Fells: Book One: The Eastern Fells* (Kentmere: Henry Marshall).

Wainwright, A. (1957) *A Pictorial Guide to the Lakeland Fells: Book Two: The Far Eastern Fells* (Kentmere: Henry Marshall).

Wainwright, A. (1958) *A Pictorial Guide to the Lakeland Fells: Book Three: The Central Fells* (Kentmere: Henry Marshall).

Wainwright, A. (1960) *A Pictorial Guide to the Lakeland Fells: Book Four: The Southern Fells* (Kentmere: Henry Marshall).

Wainwright, A. (1962) *A Pictorial Guide to the Lakeland Fells: Book Five: The Northern Fells* (Kentmere: Henry Marshall).

Wainwright, A. (1964) *A Pictorial Guide to the Lakeland Fells: Book Six: The North Western Fells* (Kendal: Westmorland Gazette).

Wainwright, A. (1966) *A Pictorial Guide to the Lakeland Fells: Book Seven: The Western Fells* (Kendal: Westmorland Gazette).

Wainwright, A. (1968) *Pennine Way Companion* (Kendal: Westmorland Gazette).

Wainwright, A. (1970) *Walks in Limestone Country* (Kendal: Westmorland Gazette).

Wallace, M. (1985) 'Mickey Mouse History: Portraying the Past at Disney World', *Radical History Review*, 32, 3–57.

Watson, B. and Scraton, S. (2001) 'Confronting Whiteness? Researching the Leisure Lives of South Asian Mothers', *Journal of Gender Studies*, 10, 265–277.

Watson, P. (2010) *The German Genius* (London: Simon and Schuster).

Weaver, A. (2011) 'The Fragmentation of Markets, Neo-Tribes, Nostalgia, and the Culture of Celebrity: The Rise of Themed Cruises', *Journal of Hospitality and Tourism Management*, 18, 54–60.

Weaver, S. (2011) *The Rhetoric of Racist Humour: Us, UK and Global Race Joking* (Farnham: Ashgate).

Wells, J. (1983) 'Me and the Devil Blues: A Study of Robert Johnson and the Music of the Rolling Stones', *Popular Music and Society*, 9, 17–24.

Wheaton, B. and Beal, B. (2003) 'Keeping It Real: Subcultural Media and the Discourses of Authenticity in Alternative Sport', *International Review for the Sociology of Sport*, 38, 155–76.

Whyte, I. (2007) 'The Lake District and the Yorkshire Dales: Refuges from the Real World?', in C. Ehland (ed) *Thinking Northern: Textures of Identity in the North of England* (Amsterdam: Rodopi).

Wilcox, R. (2005) *Why Buffy Matters* (London: IB Taurus).

Wilkinson, D. and Thelwall, M. (2010) 'Social Network Site Changes over Time: The Case of MySpace', *Journal of the American Society for Information Science and Technology*, 61, 2311–2323.

Woodward, K. (2004) 'Rumbles in the Jungle: Racialization and the Performance of Masculinity', *Leisure Studies*, 23, 5–17.

Wrong, D. (2009) *Our Turn to Eat* (London: Harper Collins).

Yarwood, R. and Charlton, C. (2009) 'Country Life? Rurality, Folk Music and Show of Hands, *Journal of Rural Studies*, 25, 194–206.

Young, R. (1995) *Colonial Desire: Hybridity in Theory, Culture and Race* (London: Routledge).

Yousman, B. (2003) 'Blackophilia and Blackophobia: White Youth, the Consumption of Rap Music, and White Supremacy', *Communication Theory*, 13, 366–391.

Yuen, F. and Pedlar, A. (2009) 'Leisure as a Context for Justice: Experiences of Ceremony for Aboriginal Women in Prison', *Journal of Leisure Research*, 41, 547–564.

Index

Acorah, Derek, 2–3
active sports tourism, 42
alcohol drinking, 142–3
American television, 73–5
athletics, 111–13
Australia, 108, 125, 128–30
authenticity, 161–2

baseball, 108
basketball, 110–11, 131–3
blackness, 4, 5, 14, 15, 38, 41, 89–90,
    131–3
blues music, 90
Bourdieu, Pierre, 19, 72, 90
boxing, 39
Buffy the Vampire Slayer, 78

class, 15, 20, 68, 84, 106, 110, 114,
    115, 124, 130, 134, 152, 160–1,
    162–4, 182–4, 196
classical music, 87–8
climbing, 29, 180–1
coffee drinking, 75, 143–4
communicative rationality, 56–8
cricket, 38, 108
Critical Race Theory, 17–18, 27–8,
    39–40
cruise ships, 164
Culture Wars, The, 80
cycling, 150

dancing, 147–8
Diamond Jubilee, The, 18
Disney vacations and theme parks,
    164–5
Dracula, 78

eating out, 152–7
eco-tourism, 44
England, 33–4, 133–6
    see also United Kingdom
ethnic food, 152–7

Fanon, Franz, 19
Fifty Shades of Grey, 82
folk music, 88, 96, 97–101
football, American, 106
football, Association (soccer), 1–2, 38,
    106, 114, 116, 133–6
Foucault, Michel, 21–3, 53, 71
Friends, 74–5

gender, 15, 21–2, 30, 37, 38, 40, 41–2,
    74, 78–9, 82–3, 106–7, 119
Gilroy, Paul, 15, 45, 70, 71

Habermas, Jurgen, 5, 23, 47–66, 71,
    140, 198–9
Hall, Stuart, 14–15, 39, 44–5, 70
Harry Potter, 67–9
heavy metal music, 97–101
hegemonic whiteness, 68, 84–5,
    101–2, 104–5, 120, 126–7, 141,
    152–3, 157, 180–1, 184, 194–6
hegemony, 19, 23–4, 29–30, 32, 45,
    50, 62, 64
heritage tourism, 34–5, 166–70
high culture, 71, 87–8, 147
horse riding, 176–7
hybridity, 15, 91–7, 162

independent travellers, 170–3
instrumental rationality, 58–63
instrumental whiteness, 63–5, 103,
    104, 120, 194, 196–7, 198–9
internet culture, 122–3, 148–52
intersectionality, 15, 21, 30, 141–2,
    198

Jewishness, 36–7, 131–3

Kenya, 143

Lake District, The, 187–91

215

Printed and bound by
CPI Group (UK) Ltd, Croydon, CR0 4YY